THE
OMNIVERSE

"Webre's compact and easy-to-read book is a brilliant compilation of solid scientific evidence for the existence of human souls that reincarnate and exist in parallel lives in many dimensions, frequently as extraterrestrials! This is a must-read for anyone who wants to fully engage with many dimensions to end absolutely any fear of abandonment, death, or the future."

BARBARA HAND CLOW, AUTHOR OF
THE ALCHEMY OF NINE DIMENSIONS,
THE PLEIADIAN AGENDA, AND
REVELATIONS OF THE RUBY CRYSTAL

"*The Omniverse* demonstrates that we have reached a critical mass of empirical evidence for the existence of intelligent species and civilizations communicating with us from beyond our earth. It will take courage to open fully to this book's vast scientific and spiritual implications. Something truly magnificent and revelatory is happening here."

EDWARD BRUCE BYNUM, PH.D., AUTHOR OF
DARK LIGHT CONSCIOUSNESS: MELANIN, SERPENT POWER
AND THE LUMINOUS MATRIX OF REALITY AND
THE AFRICAN UNCONSCIOUS

". . . heartening and affirming to all of those who have experienced multidimensional realities. For the researcher it is a book with a tantalizing perspective of what is not only possible but also that what we believed is impossible may be possible. For the seeker it's another piece of the puzzle in understanding the multilevel nature of human experiences—a wonderful journey of discovery."

MARY RODWELL, DIRECTOR OF
THE EDGAR MITCHELL FOUNDATION FOR RESEARCH INTO
EXTRATERRESTRIAL ENCOUNTERS (FREE) AND PRINCIPAL OF
ACERN (AUSTRALIAN CLOSE ENCOUNTER RESOURCE NETWORK)

"Much of the writing of Apollo astronaut Edgar Mitchell, Harvard astrophysicist Rudy Schild, Mary Rodwell, and me—the cofounders of the Foundation for Research into Extraterrestrial Encounters—has confirmed the findings of Alfred Webre: that UFOs and ETs do exist, and that they communicate with us telepathically and operate in a multiverse. We salute Webre in his efforts to illuminate humanity."

REY HERNANDEZ, ATTORNEY AND COFOUNDER OF
THE FOUNDATION FOR RESEARCH INTO
EXTRATERRESTRIAL ENCOUNTERS (FREE)

THE
OMNIVERSE

Transdimensional Intelligence, Time Travel, the Afterlife, and the Secret Colony on Mars

ALFRED LAMBREMONT WEBRE

Bear & Company
Rochester, Vermont • Toronto, Canada

Bear & Company
One Park Street
Rochester, Vermont 05767
www.BearandCompanyBooks.com

Text stock is SFI certified

Bear & Company is a division of Inner Traditions International

Library of Congress Cataloging-in-Publication Data
Names: Webre, Alfred L., 1942-
Title: The omniverse : transdimensional intelligence, time travel, the
    afterlife, and the secret colony on Mars / Alfred Lambremont Webre.
Description: Second edition. | Rochester, Vermont : Bear & Company , 2015. |
    Includes bibliographical references and index.
Identifiers: LCCN 2015015312| ISBN 9781591432159 (paperback) | ISBN
    9781591432166 (ebook)
Subjects: LCSH: Universe--Miscellanea. | Cosmology--Miscellanea. |
    Parapsychology. | Occultism. | BISAC: BODY, MIND & SPIRIT / UFOs &
    Extraterrestrials. | SCIENCE / Cosmology. | BODY, MIND & SPIRIT / New
    Thought.
Classification: LCC BF1999 .W4155 2015 | DDC 130--dc23
LC record available at http://lccn.loc.gov/2015015312

Printed and bound in the United States by Lake Book Manufacturing, Inc.
The text stock is SFI certified. The Sustainable Forestry Initiative® program
promotes sustainable forest management.

10 9 8 7 6 5 4 3 2 1
Text design by Priscilla Baker and layout by Debbie Glogover
This book was typeset in Garamond Premier Pro with Gill Sans and Helvetica
Neue for display fonts

To send correspondence to the author of this book, mail a first-class letter to the author c/o Inner Traditions • Bear & Company, One Park Street, Rochester, VT 05767, and we will forward the communication, or contact the author directly at **webre@shaw.ca** or **www.exopolitics.com**.

Contents

Acknowledgments

Given that this is a book on the multiple dimensions of the Omniverse, I would like to acknowledge the invaluable support and insights I have received in this lifetime from those dimensional persons and forces in exopolitical and spiritual dimensions that have led to the writing of this book. I would like to thank my guides, assisting hyperdimensional civilizations, assisting spiritual entities, and assisting souls, as well as the Source. Thanks also to those that have attempted to prevent the publication of this book, for you have only made it more refined.

I thank and acknowledge the incalculable help I have received over the years in the Earth dimensions from many colleagues, researchers, and friends too numerous to mention here.

I thank my editor, Aidan Kelly, who helped shape this book into a vessel for you, the reader, as he has done with so many books before as an editor with Scientific American and Jeremy Tarcher (Penguin). I also thank Nancy Yeilding for her excellent editing on this edition.

I thank my families, who bore patiently my years of research into dimensional ecology, including my parents, my siblings, my son, and my stepdaughters, and my soul spouse, Geri DeStefano-Webre, Ph.D., without whose presence this book would not have been possible.

And I thank you, the reader, who now hold yet another map into the dimensional ecology of the Omniverse in your hands. May it serve you and yours well.

The Beginning
of This Book

My journey in mapping the Omniverse came to a significant threshold in Barcelona, Spain, on a bright day in July 2009. That day Ima Sanchis, a veteran journalist with *La Vanguardia*, the leading Catalan newspaper and top newspaper in Spain, interviewed me for her full-page column *La Contra*. I was on a book tour of Spain as part of the launch of the Spanish edition of my book *Exopolitics: Politics, Government, and Law in the Universe.*"[1]

Toward the end of the interview, in a small conference room at the *La Vanguardia*'s Barcelona headquarters, Ima put down her pen, turned off her tape recorder, turned to me, and pointedly asked:

"¿Alfred, Como funciona el universo? ¿Cuál es la relación entre los extraterrestres y la reencarnación del alma humano?"

Translated into English Ima Sanchis' questions mean:

"Alfred, how does the universe function? What is the relationship between extraterrestrials and the reincarnation of the human soul?"

As Ima asked her questions, I recognized that she was asking me to address scientifically those key issues that most humans, philosophers, scientists, priests, and earthly religions have struggled to answer for millennia—without much objective scientific success. It dawned on me in that moment that Ima was pointing to a next pathway of empirical research into how intelligent civilizations in the physical multiverse relate to each other and to the larger community of intelligent civilizations of souls, spiritual beings, and Source.

I returned to Madrid after speaking in Barcelona. At my hotel in Madrid, on my way back home to Vancouver, B.C., I pondered the relationships between extraterrestrials and the soul, and scribbled notes to myself on hotel notepaper, about the basic similarities in communications and transportation between humans and extraterrestrials in this universe and between humans and souls in the afterlife. There were structural parallels between those two fields of inquiry in how information about them is organized—yet no one, as far as I knew, had ever looked explicitly at those patterns.

A second important threshold occurred shortly afterward, in August 2009, when an editor for Oxford University Press asked me to write a book applying "the principles of international law to extraterrestrial contact."

In order to write a book on that subject, I realized I would have to first define what the term "intelligent extraterrestrial civilizations" means—and at this point I discovered that no adequate typology of such civilizations had yet been developed. As I worked to develop a new typology for extraterrestrial civilizations and extraterrestrial governance bodies, I realized that it provided an evidentiary and conceptual foundation for mapping the many dimensions of the Omniverse.

A Map of the Omniverse

A map of the Omniverse? To enter into this daunting field, let us begin with some helpful definitions. First of all what *is* the Omniverse?

To understand that we need to begin with the multiverse. The *multiverse* is defined as the total of all universes, including our own, and is thought to encompass "all space, time, matter, and energy."[1] Cosmologists Richard L. Amoroso and Elizabeth A. Rauscher write:

> Generally multiverse, sometimes called meta-universe or megaverse, is the hypothetical set of multiple possible universes (including our Einstein-Hubble universe) that together comprise all of reality. . . . The structure of the multiverse, the nature of each constituent universe, and the connection between them depends on the particular multiverse hypothesis being considered by the theory. But the term *universe* is supposed to represent the entirety of all existence; however, with usages like "Mr. Tompkins Universe" or the universe of the ant, one has become accustomed to the idea of many universes, at least in the common vernacular. Interestingly, American psychologist William James first coined the term *multiverse* in 1895. In scientific circles, many disparate definitions of the multiverse exist, such as parallel universes, Bubble universes, alternate realities as in Everett's Many Worlds interpretation of quantum theory containing

every possibility, or the . . . extension of string theory known as M-theory, where our universe and others are purported to be created by collisions between membranes in an [11-dimensional] space.[2]

The total number of universes besides our own estimated to exist in the multiverse is staggering. Physicists Andrei Linde and Vitaly Vanchurin of Stanford University recently calculated "that the total number of such universes, in the simplest inflationary models, may exceed" a number that can be written as:

10 raised to the (10 raised to 10^7) power.[3]

This is a deceptively compact notation. First, 10^7 is a 1 with 7 zeroes after it, that is, 10,000,000, or ten million. Next, 10 raised to the 10-millionth power, that is, $10^{10,000,000}$, is a 1 with 10 million zeroes after it. Written out with 6 zeroes to the inch, it would stretch for about 26 miles. But the next step, raising 10 to the power of that 26-mile number, generates a number so large that we cannot name it, let alone write it out. It would stretch for at least 260 million miles. Linde and Vanchurin also said, "This humongous number is strongly model dependent and may change when one uses different definitions of what is the boundary of eternal inflation."[4]

To the concept of the multiverse, we need to add that the universes of the multiverse are here collectively termed the *exopolitical dimensions,* which highlights their function as a base for intelligent civilizations. The science of exopolitics is the study of the relations between intelligent civilizations in the multiverse.

Finally, to arrive at a comprehensive understanding of what we mean by *Omniverse,* one more aspect needs to be considered: the spiritual dimensions, which include the intelligent civilizations of souls, spiritual beings, and Source (God). The Omniverse thus expands on the conventional scientific definition of the multiverse as consisting solely of parallel physical universes of time, space, energy, and matter.

The spiritual dimension of the Omniverse provides the energy (energy that scientists such as Lawrence M. Krauss, author of *A Universe from Nothing,* cannot now account for[5]) needed for the creation and maintenance of each physical universe in the multiverse. The Omniverse is the sum of the multiverse plus the spiritual dimensions—in other words, all that is.

Our understanding of the Omniverse can be expressed as two fundamental equations:

Multiverse (exopolitical dimensions) = sum of all physical universes

Omniverse = multiverse (exopolitical dimensions) + spiritual dimensions (intelligent civilizations of souls, spiritual beings, and Source/God)

These equations are the basis for the hypothesis I have termed "the dimensional ecology of the Omniverse." Wikipedia defines *ecology* as "the scientific analysis and study of interactions among organisms and their environment."[6] *Dimensional ecology,* then, is "the analysis and study of interactions among organisms and their environment in interlocking dimensions of existence, including the universes of the multiverse and the spiritual dimensions."

Hypothesis

Dimensional Ecology of the Omniverse

This hypothesis is based on the understanding that *dimensionality*—the ability of intelligence to organize itself via dimensions (discrete bands of conscious energy)—is a key criterion by which the Omniverse is designed, in both the spiritual dimensions and the exopolitical dimensions. According to this hypothesis, our human civilization on Earth is but one of many intelligent civilizations based in parallel dimensions and universes in the multiverse, along with the intelligent civilizations of souls and spiritual beings, as well as Source (God), in the spiritual dimensions. By *souls* I mean individuated, nonlocal, conscious, intelligent entities that are based in the dimensions of the afterlife. *Spiritual beings* are similarly individuated, nonlocal, conscious, and intelligent entities that are based in alternative locations of the spiritual dimensions, are created by a different process than souls, and have a different evolutionary spiritual destiny. Typically spiritual beings do not reincarnate within the universes of the multiverse, but rather tend and minister to souls that are in some type of process, such as reincarnating.

One way of stating the hypothesis is to say that we Earthling humans live within an Omniverse that encompasses both:

Exopolitical dimensions, consisting of our known physical universe and multiple parallel universes within which are based extraterrestrial and hyperdimensional* intelligent civilizations

Spiritual dimensions, consisting of the "afterlife" or "interlife"† and other spiritual dimensions within which are based the intelligent civilizations of souls and spiritual beings as well as a defined Source (God)

A hypothesis is, by accepted definition, "an educated guess, based on observation. Usually a hypothesis can be supported or refuted through experimentation or more observation. A hypothesis can be disproven, but not proven to be true."[7] The dimensional ecology hypothesis is a guide to intelligent life in the Omniverse. It is an edu-

---

Hyperdimensional civilizations are intelligent civilizations based on dimensions in our universe parallel to our own time-space Earth dimension or in universes parallel to our universe.

†Our research has found evidence (presented in chapter 6) for the existence of reincarnation as a regular feature of intelligent life in the universes of the multiverse. Hence, we use the term *interlife* interchangeably with *afterlife* to designate the period during which a soul locates fully in the spiritual dimensions between incarnations in a universe.

cated guess, based on observations such as witness testimony, data, and cases of how, where, and why intelligent life exists in all of the Omniverse. It is supported by the empirical scientific method, using the jurisprudential law of evidence that employs witness, documentary, and experimental evidence.

In mapping the Omniverse we will employ both scientific evidence and legal evidence. According to one source, "Types of legal evidence include testimony, documentary evidence, and physical evidence. . . . Scientific evidence consists of observations and experimental results that serve to support, refute, or modify a scientific hypothesis or theory, when collected and interpreted in accordance with the scientific method."[8] For example, we will evaluate reliable eyewitness accounts from project members of U.S. government secret Mars programs (legal evidence). Likewise, we will also draw support for our hypotheses from the results of cases that used hypnotic regression under standard laboratory protocols to achieve replicable soul memories of the afterlife (scientific evidence).

Part 1 of the book, "The Omniverse and Its Inhabitants," lays the groundwork for what follows. Chapter 1 explores current scientific estimates of how many universes exist in our multiverse and reviews the best empirically based estimates of the number of intelligent civilizations in our galaxy, universe, and multiverse. It also introduces the intelligent civilizations of souls, of spiritual beings, and of Source (God) in the spiritual dimensions, thus providing what string theory has not been able to: a framework for understanding the energy that human scientists cannot now account for in the creation and maintenance of each physical universe in the multiverse. Making it clear that the dimensional ecology hypothesis fulfills science's current need for a more comprehensive hypothesis, the chapter goes on to detail the data sources and methods of exopolitics and parapsychology, the two key sciences of this hypothesis.

The dimension-based typology of intelligent civilizations in the multiverse is introduced in chapter 2. Unlike existing typologies this typology is based on how intelligent civilizations describe themselves, as being based in a specific dimension, "density," or energy frequency—which is a civilization's most fundamental typological, locational, and developmental criterion. This new typology establishes dimension-based criteria for extraterrestrial and interdimensional civilizations and extraterrestrial governance bodies, based on recent whistle-blower, direct witness, and documentary evidence.

Concurrent with the dimension in which a specific intelligent species or civilization is based, an intelligent species or civilization can also be typed by "exophenotypology," which is the typology or classification of extraterrestrials based on the observable characteristics of their physical appearance. Our map of the dimensional ecology explores the humanoid, Grey, reptilian, insectoid, and other exophenotypologies of intelligent civilizations in the multiverse.

In part 2, "Exopolitical Dimensions of the Omniverse," chapters 3, 4, and 5, our map specifically focuses on the multiverse, exploring the intelligent civilizations based in the physical universes of time, space, energy, and matter. Chapter 3 explores a case study of the evidence for an intelligent civilization based in the time-space dimension on a planet in our own solar system, in this case, the existence of three humanoid exophenotypes on Mars. This conclusion is the result of empirical prima facie evidence, evidence that is sufficient, on the face of it, to support a claim. It includes the evidence of eyewitnesses and of documentary (photographic) evidence, including that from the NASA Mars rovers.

Chapter 4 explores the persuasive empirical evidence for how hyperdimensional intelligent civilizations operate within the multiverse. Whether based in higher dimensions of our own solar system, our galaxy (the Milky Way), other galaxies of our universe, or universes parallel to our own universe in the multiverse, these hyperdimensional intelligent civilizations may use technologically advanced interdimensional transport as well as advanced consciousness technologies when teleporting into our known physical universe or our Earth time-space dimension.

Chapter 5 explores the prima facie evidence for the existence of intelligent civilizations' governance authorities in the universes of the multiverse, which have jurisdiction over defined dimensional territories, such as the Milky Way galaxy in our universe. Two types of evidence are acceptable to the science of exopolitics. The first is replicable scientific remote viewing. The second is eyewitness and documentary evidence, consisting of eyewitness contactee and telepathic interaction with representatives and intermediaries of reported intelligent civilizations' governance authorities.

Part 3, including chapters 6 through 8, explores the "Spiritual Dimensions of the Omniverse." The parallel "visible" universes (or the multiverse) are more aptly termed the exopolitical dimensions, since they are where the intelligent civilizations of souls that are based in the spiritual dimensions incarnate in a variety of intelligent creatures for the purpose of moral and soul development. The communications of souls based in the interlife dimensions to Earthlings in the time-space hologram provide the evidence for the existence of the spiritual dimensions.

Chapter 6 explores the prima facie evidence for the interlife (afterlife) dimensions. It is highly meaningful for the dimensional ecology of the Omniverse hypothesis that the two Italian Catholic priests, Fathers Pellegrino Ernetti and Augustino Gemelli, who developed instrumental transcommunication (ITC) as a technology for exploring the interlife in the spiritual dimensions also developed chronovision, a time-travel technology for exploring timelines in the multiverse.

Chapter 7 explores the prevailing evidence, absent evidence to the contrary, for the intelligent civilizations of souls. These individuated, nonlocal, conscious, intelligent entities based in the interlife dimensions are holographic fragments of the original Source or Creator. Replicable data report that souls are created in a process that results in each one being a holographic "egg of Light" drawn from the original Source.[9] Soul memories of the interlife obtained from hypnotic regression support the dimensional ecology of the Omniverse hypothesis, since they provide detailed information regarding the dimensional interactions of the intelligent civilizations of souls, of spiritual beings, and of Source (God) with the universes and intelligent civilizations of the multiverse.

The central role played by the intelligent civilizations of souls in the creation and maintenance of the universes of time, space, energy, and matter of the multiverse is also described in chapter 7. There is evidence that souls are involved in the creation of galactic matter, stars, and planets and perhaps life as well.

Chapter 8 addresses Grey hyperdimensional control over soul reincarnation procedures. There is empirical evidence that human souls have been aboard some hyperdimensional Grey spacecraft, undergoing complex soul reincarnation operations that are carried out by the Grey extraterrestrials. We explore a case study that demonstrates that souls and specific Grey hyperdimensionals navigate the dimensional ecology of the multiverse as interdimensional entities in cooperative relationships.

Our map of the Omniverse concludes with a presentation of the ten top implications for a positive human future that can be drawn from research into the science of the Omniverse, encouraging further exploration of the dimensional ecology of the Omniverse.

PART 1

The Omniverse and Its Inhabitants

1

Understanding the Omniverse

Current science has been making great strides in understanding our universe. For example, working with the Baryon Oscillation Spectroscopic Survey (BOSS), researchers have now developed an "ultraprecise galaxy map" that has measured the distance to galaxies in our universe more than 6 billion light years away to within 1 percent accuracy. A report on the map stated:

> "There are not many things in our daily lives that we know to 1 percent accuracy," David Schlegel, a physicist at Lawrence Berkeley National Laboratory and the principal investigator of BOSS, said in a statement. "I now know the size of the universe better than I know the size of my house."
>
> The new results, presented by Schlegel and his colleagues [on January 8, 2014] at the 223rd meeting of the American Astronomical Society, also provide one of the best-ever determinations of the curvature of space, researchers said. In short, the universe appears to be quite "flat," meaning that its shape can be described well by Euclidean geometry, in which straight lines are parallel and the angles in a triangle add up to 180 degrees.

"One of the reasons we care is that a flat universe has implications for whether the universe is infinite," Schlegel said. "That means—while we can't say with certainty that it will never come to an end—it's likely the universe extends forever in space and will go on forever in time. Our results are consistent with an infinite universe."[1]

This tentative finding that our universe is infinite is congruent with Linde and Vanchurin's finding that there are a "humongous" number of universes in the multiverse, since universes that extend forever in space and go on forever in time can coexist in parallel with each other.

Truly comprehending infinity, whether of one universe or an incalculable number of them, is difficult enough. But even science is reaching its limits when it comes to understanding the many dimensions of our Omniverse—which contains the totality of all the universes of time, space, energy, and matter, as well as all of the intelligent civilizations in the multiverse *in addition to* the intelligent civilizations of souls in the interlife (afterlife) dimensions, and of spiritual beings and Source (God) in the spiritual dimensions.

Awareness of the essential elements, if not the details, of the dimensions of the Omniverse has existed in diverse and fragmentary forms for millennia in our earthly religions, in our sacred and occult texts, and in the group and individual beliefs of human societies throughout history. The multidimensional Omniverse is thus both a scientific hypothesis and a truth that many Earthling humans are familiar with intuitively or have learned during their lives in their families, schools, religions, reading, or philosophizing.

Humans are familiar with many concepts included in the dimensional ecology hypothesis, such as afterlife, souls, and Source (God), although they have widely differing opinions, depending on race, nationality, gender, and education, about whether such things actually exist and in what specific form. Many assertions about specific aspects, such as life after death, are set out in the religious beliefs and sacred texts that organized religions have relied on over millennia to win followers. However, important details about, for example, the nature of the afterlife as described by various religions, are sometimes factually incorrect when compared to replicable, scientifically derived knowledge about what happens after physical death. In what follows we will be comparing traditional beliefs with what we can now know from scientific research.

Intelligent Civilizations in Our Universe and Multiverse

One German supercomputer simulation estimates that there are "500 billion galaxies in our universe."[2] Astronomers now estimate there are 100 billion habitable Earth-like planets in our Milky Way galaxy and 50 sextillion habitable Earth-like planets in our particular universe.[3] A conservative estimate of the number of communicating intelligent civilizations in our universe is 100 billion (100,000,000,000). This estimate is based on the 1960 Drake equation, which assumes that there are only twelve communicating intelligent civilizations in our Milky Way galaxy, out of the estimated 100 billion habitable Earth-like planets.[4]

Then how many intelligent civilizations may there be in the multiverse? If we multiply the Drake equation–based estimate of a hundred billion communicating intelligent civilizations in our universe by Linde and Vanchurin's calculation of the number of universes in the multiverse, we arrive at the number of intelligent civilizations in the multiverse as being 100,000,000,000 times that 260-million-mile-long number. It is not physically possible to actually write that number out fully.

World Public Opinion and
Intelligent Life in the Multiverse

World public opinion is congruent with the recent science-based estimates that there are 100 billion communicating intelligent civilizations in our universe, and an even more "humongous" number of communicating intelligent civilizations in the multiverse.

The possibility that we live in a populated cosmos is conventionally thought to be controversial and esoteric. However, public opinion around the world is divided as to whether we are alone in the cosmos. This is despite the fact that the world public has been quarantined from real knowledge about the actual role of non-Earth intelligent civilizations on Earth. Instead of public education about an extraterrestrial presence and Earth's history in the galaxy, governments have knowingly fed the world public a steady diet of disinformation and brainwashing about Earth's dealings with intelligent civilizations. Ever since the 1953 U.S. Central Intelligence Agency Robertson Panel, the facts of intelligent civilizations and their visitations to Earth have been classified and off limits for civil society.[5] Consequently, world public opinion about the presence of extraterrestrial and hyperdimensional civilizations in Earth's environment is divided. Nevertheless, there is a core of world public opinion and of public opinion in specific nations that accepts that humanity coexists in a cosmos populated by other intelligent civilizations.

One 2013 public-opinion poll of 5,886 U.S. adult residents found that "37 percent affirmed a belief in the existence of extraterrestrial life, 21 percent denied such a belief, and 42 percent were uncertain, responding 'I'm not sure.'"[6] A 2010 poll by the French market-research company Ipsos of world public opinion on extraterrestrials found that "one in five (20 percent) of presumably human adults surveyed in 22 countries (representing 75 percent of the world's GDP) say they believe that alien beings have come down to Earth and walk amongst us in our communities disguised as 'us'." People in India (45 percent) and China (42 percent) are most likely to believe that extraterrestrials are visiting Earth.[7]

Intelligent Civilizations
in the Spiritual Dimensions

The universes of the multiverse are not the only dimensions where intelligent civilizations are based. There is prima facie evidence of intelligent civilizations that are based in dimensions that are outside the multiverse. These are what we term the "spiritual dimensions."

One important database of empirical evidence for the existence of intelligent civilizations in the spiritual dimensions is derived from more than 7,000 cases of hypnotic regressions in which soul memories of the interlife (or afterlife) have been retrieved, developed according to a standard laboratory protocol by Michael Newton, Ph.D.[8] This database contains evidence for the intelligent civilizations of souls, for the intelligent civilizations of spiritual beings, and for the original Source (God), the creator of the spiritual dimensions.

Humanity's basic opinions and beliefs about the interlife (afterlife), soul, and Source (God) are largely formed both by religions and by science. Major religions, such as Christianity, Islam, Hinduism, Buddhism, Taoism, Confucianism, Shintoism, Sikhism, and Judaism, and a dozen medium-sized religions have over 6.5 billion followers, or 90 percent of the total world population of approximately 7.2 billion people.

Religions are not scientific organizations that reach their conclusions about the interlife (afterlife), soul, and Source (God) on the basis of laboratory protocols and empirical science. Religions are historical, social, and political organizations whose missions and teaching about the nature of the interlife (afterlife), soul, and Source (God) are generally based on sacred texts. The content of religions' belief systems about such core issues may differ in detail from religion to religion.

Currently, there is a schism within science as to whether the interlife (afterlife), soul, and Source (God) actually exist. Academic science largely prohibits research into and teaching about parapsychology and exopolitics, two sciences necessary for exploring the empirical evidence for such existence.

World Public Opinion and the Spiritual Dimensions

According to a 2011 Ipsos poll taken in twenty-three nations among 18,829 adults, "one half (51 percent) of global citizens definitely believe in a 'divine entity' compared to 18 percent who don't and 17 percent who just aren't sure." The Ipsos poll also found "Similarly, half (51 percent) believe in some kind of afterlife, while the remaining half believe they will either just 'cease to exist' (23 percent) or simply 'don't know' (26 percent) about a hereafter. Seven percent of respondents believe in reincarnation."[9] A substantial core of world public opinion thus has a view of reality that is congruent with the prevailing replicable empirical evidence for an interlife (afterlife), intelligent civilizations of souls, civilizations of spiritual beings, and a Source (God) or Creator.

In the United States of America, a December 2013 Harris poll showed that a substantial majority of Americans hold positive beliefs about the existence of God and the essentials of the spiritual dimensions, although the percentage of Americans holding these beliefs has declined somewhat in recent years. The poll finds that 74 percent of U.S. adults do believe in God, but this represents a decline from previous years, when 82 percent expressed a belief in God in 2005, 2007, and 2009. Belief in the survival of the soul after death has declined from 69 to 64 percent in recent Harris polls, but 24 percent of Americans believe in reincarnation.[10]

The Harris poll addresses generational and political divides.

Echo Boomers are less likely than their counterparts in all older gen-
erations to express belief in God (64 percent Echo Boomers, 75 percent Gen Xers, 81 percent Baby Boomers, 83 percent Matures). . . .
On the other end of the generational spectrum, Matures are far less
likely than any other generation to express belief in ghosts (44 percent Echo Boomers, 46 percent Gen Xers, 46 percent Baby Boomers,
24 percent Matures), witches (27 percent, 29 percent, 28 percent and
18 percent, respectively), and reincarnation (27 percent, 25 percent,
23 percent, and 13 percent, respectively). Turning to the political spectrum, Democrats and Independents show similar levels of
belief in most of the tested concepts, with Republicans consistently
more likely than either group to express belief in those concepts
aligned with the Judeo-Christian belief system; Republicans are
less likely than either group to express belief in Darwin's theory of
evolution (36 percent Republicans, 52 percent Democrats, 51 percent Independents).[11]

The Harris poll notes that absolute certainty that there is a God is
down from what it was ten years ago.

In a separate line of questioning, focused on Americans' degree of certainty that there is or is not a God, two-thirds of Americans (68 percent) indicate being either absolutely or somewhat certain that there is
a God, while 54 percent specify being absolutely certain; these figures
represent drops of 11 and 12 percentage points, respectively, from 2003
testing, where combined certainty was at 79 percent and absolute certainty was at 66 percent. Meanwhile, combined belief that there is no
God (16 percent) and uncertainty as to whether or not there is a God
(also 16 percent) are both up from 2003 findings (when these levels
were 9 percent and 12 percent, respectively). Outside of specific religious samples, the groups most likely to be absolutely certain there is
a God include blacks (70 percent), Republicans (65 percent), Matures

(62 percent) and Baby Boomers (60 percent), Southerners (61 percent) and Midwesterners (58 percent), and those with a high school education or less (60 percent).[12]

As to the gender of God, the Harris poll notes,

There continues to be no consensus as to whether God is a man or a woman. Nearly 4 in 10 Americans (39 percent) think He is male, while only 1 percent of U.S. adults believe She is female. However, notable minorities believe God is neither male nor female (31 percent) or both male and female (10 percent). Women, perhaps surprisingly, are more likely than men to believe that God is male (43 percent women, 34 percent men), while men are more likely to believe that God is neither male nor female (34 percent men, 28 percent women). As to God's control over the Earth, there is also a continuing—and increasing—lack of consensus as to how much control, if any, God has over what happens on Earth. A 37 percent plurality of Americans (including 52 percent of Catholics) believes that God observes but does not control what happens on Earth, down considerably from 2003, when half of Americans (50 percent) expressed this belief. Just under three in ten (29 percent) Americans, including majorities of those who self-identify as very religious (60 percent) and/or born-again Christians (56 percent), believe that God controls what happens on Earth.[13]

As to religious texts as "Word of God," the Harris poll notes,

Just under half of Americans believe that all or most of the Old Testament (49 percent) and the New Testament (48 percent) are the "Word of God," representing declines of six percentage points each from 2008 findings. Just under two in ten Americans (19 percent) describe themselves as being "very" religious, with an additional four in ten (40 percent) describing themselves as being "somewhat" religious (40 percent, down from 49 percent in 2007). Nearly one-fourth of Americans (23 percent) identify themselves as "not at all" religious—a figure that has nearly doubled since 2007, when it was at 12 percent.[14]

The Literature on Dimensional Consciousness

Alternative expressions that are congruent with the dimensional ecology of the Omniverse hypothesis can be found in the literature on dimensional consciousness, in which levels of consciousness are understood to consist of bounded ranges and frequencies of energy. Diverse levels of consciousness and life are unique to each dimension, such as the dimensions of time and light in our universe, which Earthling human science is only beginning to explore. Dimensional levels of consciousness in our universe also interface with the exopolitical and spiritual dimensions of the Omniverse.

Cosmologists Barbara Hand Clow and Gerry Clow have explored nine dimensions of consciousness specific to our universe, galaxy, and our Earth. They state that: the first and second dimensions are levels within our three-dimensional Earth (3D), from core to surface; the fourth dimension is the realm of collective thought that bridges the physical and unseen worlds; and the fifth through ninth dimensions are celestial. They suppose that most scientists will be unable to imagine the unseen realms that reach beyond the fourth dimension except by means of mathematics, such as topology and superstring theory.[15]

Science's Need for a New Hypothesis

In order to understand the Omniverse, we have to go beyond the limitations of current science. Advanced conceptual physicists such as Professor Amit Goswami have argued that contemporary science's assumption that "only matter—consisting of atoms or, ultimately, elementary particles—is real" is inadequate and that a new hypothesis of reality is required.

Professor Goswami writes:

> We cannot connect quantum physics with experimental data without using some schema of interpretation, and interpretation depends on the philosophy we bring to bear on the data. The philosophy that has dominated science for centuries (physical, or material, realism) assumes that only matter—consisting of atoms or, ultimately, elementary particles—is real; all else are secondary phenomena of matter, just a dance of the constituent atoms. This worldview is called realism because objects are assumed to be real and independent of subjects, us, or of how we observe them. The notion, however, that all things are made of atoms is an unproven assumption; it is not based on any direct evidence for all things. When the new physics confronts us with a situation that seems paradoxical from the perspective of material realism, we tend to overlook the possibility that the paradoxes may be arising because of the falsity of our unproven assumption. (We tend to forget that a long-held assumption does not thereby become a fact, and we often even resent being reminded.) *Many physicists today suspect that something is wrong with material realism but are afraid to rock the boat that has served them so well for so long. They do not realize that their boat is drifting and needs new navigation under a new worldview.*[16] [Emphasis added]

Critiquing Professor Goswami's work, Professor Facundo Bromberg, a computer information scientist, writes:

> Science is experiencing one of the most difficult challenges of all history: to explain humans' most basic capacity, the mental. It was for

long thought that human mental abilities were insurmountably difficult to understand and therefore [consigned to the] transcendental
and disconnected world and relegated to religion. But the advent of
the paradoxes in our understanding of the physical world, plus . . .
research in artificial intelligence and cognitive science faced science
with the necessity of dealing with the mysteries of the mind. Every
single discipline in humanity has something to say on the issue: science, religion, art, poetry, etc. But none has yet the capacity of proving its assumptions.[17]

The dimensional ecology of the Omniverse hypothesis and the
replicable presumptive evidence that supports it provide the necessary
tools for a new hypothesis of reality, as well as for the desired "new
navigation under a new worldview," the need for which Professor
Goswami has identified.

Exopolitics and Parapsychology
as Key Sciences of the Omniverse

Exploring and understanding the dimensional ecology of the
Omniverse is facilitated by the correlation of parallel research in two
key sciences: exopolitics and parapsychology. Parapsychology is the
science that studies psi consciousness, including telepathy, survival of
bodily death, the soul, and reincarnation. Exopolitics is the science that
studies relations among intelligent civilizations in the multiverse.

THE SCIENCE OF EXOPOLITICS

The model of exopolitics was first publicly presented in my June
2000 lecture at a conference at the University of Wyoming chaired
by Professor Leo Sprinkle.[18] The science of exopolitics formally began
with my 1999 book, *Exopolitics: A Decade of Contact,* which was published online in the year 2000 as a free e-book to seed a new paradigm

of thought for Earthling humankind.[19] Disclosure of, and governmental and public acknowledgement and understanding of, an extraterrestrial and multidimensional presence on Earth has accelerated since 2000. *Exopolitics* was nominated for 2005 Word of the Year.[20]

As a science exopolitics uses tools familiar to other sciences. These include the philosophy of science, a research database of standard classifications, and the law of evidence as a methodology for evaluating witness and documentary evidence. Exopolitics uses the concepts and structures of cosmology and ontology to frame essential questions.

As a branch of philosophy and science, cosmology addresses the nature of the Omniverse, asking questions about the functions, components, and purpose of the Omniverse. Ontology, as a branch of philosophy, addresses issues of being, existence, and reality. As a discipline ontology grapples with the existence or nonexistence of entities such as intelligent civilizations in the Omniverse. Exopolitics develops typologies of intelligent civilizations in the dimensions and parallel universes of the physical universes of the multiverse in order to research, map, explain, and understand the dynamics and interrelations of these intelligent civilizations.

The exopolitics model, like the dimensional ecology hypothesis, is the obverse of the twentieth-century canon of science, which holds that all intelligent life ends at Earth's orbit. The penalty for any professional not toeing this line is probable loss of professional status and possible harassment by government authorities. The universe-centric exopolitics model holds instead that we live in a highly populated universe in a multiverse, filled with intelligent, evolving civilizations operating under universal laws.

We, on Earth, are just becoming aware that we live in a populated universe and multiverse. As society shifts from a twentieth-century model of canonical science to a twenty-first-century model of science, exopolitics is the empirically based science that is willing to examine the evidence that the Omniverse is filled with intelligent, evolving civilizations. Like all models the exopolitics model provides a path for minds to go where they have not ventured before. The exopolitics model provides the interface between Earthling human, extraterrestrial, and interdimensional intelligent societies in the Omniverse. In this book we use the terms interdimensional and hyperdimensional interchangeably, except that hyperdimensional has the connotation of being based in a dimension above our third time-space dimension, that is the fourth dimension.

The science of exopolitics also addresses questions raised by empirical evidence and replicable scientific knowledge about intelligent civilizations in the Omniverse. Core exopolitical questions include: Are there non-Earthling intelligent civilizations in the universes of the multiverse, and how can we know and test this? What forms do intelligent civilizations take in the Omniverse? How can we develop a typology of intelligent civilizations in the universes of the multiverse?

EXOPOLITICAL RESEARCH DATA

Primary databases for exopolitical research and activity can be derived using the scientific method from a wide variety of sources, including at least the following formal categories of primary research data:[21]

1. Reports from whistle-blowers from inside secret extraterrestrial liaison programs
2. Documentary evidence (including photographs, video, radar reports, etc.) from government and official sources
3. Replicable information derived through technology-based, out-of-body experiences, Montauk chair device, chronovisor, teleporter, stargate, plasma chamber, and jump room
4. Replicable scientific remote viewing using standard laboratory protocols

Omniverse Technologies

The *chronovisor* is a device invented by Benedictine priests Pellegrino Ernetti and Augustino Gemelli, which consists of a series of antennas capable of receiving communication from the intelligences of specific deceased humans.

Stargates, or "time doors," are natural passageways that exist throughout the universe. They allow travel to different times and places and are routinely used by souls as a basic form of travel.

The *Montauk chair,* according to former chrononaut Andrew D. Basiago, is "a magnetic transducer that amplifies human psychic performance so that the person in the chair experiences or sees moments in his or her own subjective future."[22] Mr. Basiago describes teleporting—via a plasma confinement chamber developed by scientist Sterling Colgate—and then going through a wormhole through the quantum plenum to arrive at a destination in past time-space coordinates.

Other quantum access technologies for accessing the time-space hologram include the *teleporter* and the *jump room.* The teleporter set up a radiant energy field through which the teleportee entered and was propagated through a vortal tunnel to a destination in the time-space hologram. The jump room was a technology reportedly gifted to the U.S. government by a specific species of Grey extraterrestrials and consisting of an enclosed elevator space that on command morphed into a unit that teleported from one of several locations in the United States to secret U.S. landing bases on Mars.[23]

5. Reports from voluntary conscious physical contactees

6. Reports from involuntary semiconscious physical contactees (abductees)

7. Reports from voluntary semiconscious alter-physical contactees, in which the contactee's body for example will remain in bed,

while his or her etheric body and soul are teleported aboard a hyperdimensional spacecraft

8. Reports from voluntary psychic contactees (channelers and visionaries)
9. Reports from neutral psychic contactees
10. Reports from superficial excited witnesses and sightings reports
11. Reports from astute debriefers, debunkers, and interpreters
12. Alien artifacts
13. Independent archaeology, including exoarchaeology
14. Reports from shamans and occult societies
15. Science fiction
16. Revelations authorized by universe, galactic, or regional governance authorities

THE SCIENCE OF PARAPSYCHOLOGY

Parapsychology employs the empirical scientific method to study Omniverse phenomena such as psychic capacity, reincarnation of the soul, life after death, the interlife, the human soul, spiritual beings, the spiritual dimensions of the Omniverse, and God or Source. Parapsychological researchers include J. B. Rhine (psychic capacity), Chris Carter, Russell Targ (remote viewing), Ian Stevenson (reincarnation), Raymond A. Moody (near-death experiences), and Michael Newton (the interlife).

Core cosmological questions for parapsychology include:

✸ How can we know and test whether the Omniverse includes a dimension of the interlife, or life after bodily death?

✸ How does the typology of intelligent life in the interlife dimension— the life of the soul—relate to a typology of intelligent life in the physical dimensions of the Omniverse—the life of the body?

✸ What does an integrated model of science-based knowable, provable intelligent life in the Omniverse look like?

The Time-Space
Hologram of Our Universe

In "Simulations back up theory that Universe is a hologram," Ron Cowen writes, "At a black hole, Albert Einstein's theory of gravity apparently clashes with quantum physics, but that conflict could be solved if the Universe were a holographic projection."[24] In fact, the physical universe of time, space, energy, and matter that we Earthling humans inhabit *is* holographic.[25]

The *holographic principle* was first proposed by theoretical physicists Gerard 't Hooft and Leonard Susskind back in 1993. PBS's Kate Becker explains:

> The holographic principle, simply put, is the idea that our three-dimensional reality is a projection of information stored on a distant, two-dimensional surface. Like the emblem on your credit card, the two-dimensional surface holds all the information you need to describe a three-dimensional object—in this case, our universe.[26]

Writer Danielle Wiener-Bronner further explains, "dimensions [like the time-space continuum] in our universe are flat and the hologram expressing a dimension is multi-dimensional, so that 3rd dimensional time-space human beings like ourselves who exist inside the time-space hologram perceive the Earth and the planets in our solar system as round, which they are inside the hologram."[27]

New research from scientists at the Vienna University of Technology concludes that the holographic principle is possible in the context of a mostly flat space-time continuum. The time-space hologram within which our Earth human civilization exists is created and maintained by the spiritual dimension composed of the intelligent civilizations of souls and of spiritual beings, along with Source (God). They are the source of the totality of the universes of time, space, matter, and energy in the exopolitical dimensions (the multiverse).

Exploring the Many Dimensions
of the Omniverse

Telepathy, time travel, and teleportation are methods by which intelligent civilizations navigate both the exopolitical dimensions (the multiverse) and the spiritual dimensions of the Omniverse. Telepathy is nonlocal mind-to-mind communication. Time travel consists of movement across more than one timeline. Teleportation consists of point-to-point movement across a single timeline. Teleportation can occur within dimensions, such as the time-space hologram that Earthling humans inhabit, or between dimensions and parallel universes of the exopolitical dimensions.

THE DIMENSIONAL COMMONALITY OF TELEPATHY
AND TELEPORTATION

The common factors of telepathy and teleportation seem to reveal fundamental similarities between communication and transportation among humans, intelligent civilizations of extraterrestrials, and intelligent civilizations of souls.

Communications between mediums and representatives of soul civilizations in the interlife, as exemplified by the many apparently genuine reported cases of mediumistic communications with persons in the afterlife, are telepathic. Communications among human contactees and representatives of hyperdimensional intelligent civilizations also can be telepathic, as exemplified by the reported extraterrestrial contactee cases.

A common mode of transportation among human contactees and representatives of intelligent civilizations can be teleportation, as when human abductees are teleported to a waiting spacecraft by a hyperdimensional civilization. Transportation between the time-space dimension in our physical universe and the afterlife also appears to be teleportation. For example, parapsychological research reports that when an individual soul departs from a physical body that has died, the soul teleports to an interdimensional portal between our time-space dimension and the afterlife dimension.

TELEPORTATION AND TIME TRAVEL, NATURAL AND TECHNOLOGICAL

Teleportation and time travel can happen through naturally occurring stargates or interdimensional portals within both the exopolitical dimensions (the multiverse) and the spiritual dimensions.[28] Empirical studies show that, at bodily death in the time-space hologram, a human soul has the option of teleporting to a dimension within the spiritual dimensions called the afterlife (or interlife) through a natural dimensional portal from the time-space hologram in the exopolitical dimension to the interlife in the spiritual dimensions.[29]

Teleportation and time travel in the time-space hologram can also occur with advanced technology. We now know from former chrononaut Andrew D. Basiago, a project whistle-blower, that the U.S. Department of Defense's Defense Advanced Research Projects Agency's (DARPA) Project Pegasus had developed teleportation and time travel in the period 1969–72.

Project Pegasus was a secret scientific program within DARPA in liaison with the Central Intelligence Agency and under Department of Defense attaché Donald H. Rumsfeld of President Richard M. Nixon's cabinet. The program goals were to develop working applications for quantum access through time travel across multiple timelines to time-space coordinates in the past and the future, and through teleportation to multiple time-space coordinates along a single timeline.

Mr. Basiago has revealed that between 1969 and 1972, as a child participant in Project Pegasus, he viewed past and future events through a device known as a chronovisor. In addition, he teleported back and forth across the country in vortal tunnels (time-space energy tunnels) opened in time-space via Tesla-based teleporters located at the Curtiss-Wright Aeronautical Company facility in Wood Ridge, NJ, and at the Sandia National Laboratory in Sandia, NM.[30]

Secret U.S. government time-travel technology has generally been used for remote sensing in the time-space hologram. Basiago reported publicly that he saw a physical copy of my 2005 book, *Exopolitics: Politics, Government, and Law in the Universe,* in 1971 in the presence of two other witnesses when DARPA's Project Pegasus used advanced time-travel technology to teleport the book from at least 2005 back in time to 1971. As corroboration, during 1971–72, I was General Counsel of New York City's Environmental Protection Administration. At that time I was invited to give an environmental talk to approximately fifty officials. After 2005 I learned that the fifty officials I had spoken to in 1971–72 were in fact DARPA, U.S. Department of Defense, and CIA officials who had been briefed on my future book *Exopolitics*. This time-travel surveillance intelligence mattered to the U.S. Department of Defense and the CIA.

Challenges to
Understanding the Omniverse

In addition to the inherent challenges in understanding the Omniverse, related to its size and complexity, we also have to contend with various forms of purposeful interference.

For example, in 1977, while working as a futurist at the Stanford Research Institute, I became the director of the proposed Jimmy Carter White House Extraterrestrial Communication Study.[31] That exposed me to the knowledge of how time travel and teleportation have been weaponized rather than used as technologies for the public benefit, such as public teleportation. Time travel in the form of time loops has been used to sequester military secrets and secret military installations.

For example, in 1982 the U.S. government would use time travel to hide military secrets from its Cold War arch foe the Soviet Union by time traveling the documents containing the secrets to a secret location in time-space in 1995, to which the Soviet Union did not have access in 1982 as the Soviet military reportedly did not at that time have time-travel technology.

Time travel has been used for political surveillance and political control of the human population via time-travel pre-identification of future U.S. presidents. Secret U.S. government teleportation has been used to teleport project participants to government project sites. According to Andrew D. Basiago, in the early 1970s, DOD time-travel and teleportation liaison Donald Rumsfeld decreed that teleportation would be used to deploy troops to the battlefield, and not released to the public despite its potential application as a global nonpolluting transportation technology and substitute for the petroleum-driven transportation infrastructure.

Another form of challenge to exopolitical research is that of evaluating content on cosmological concepts that is derived from revelations authorized by universe, galactic, or regional governance authorities. Some sources of interdimensional information seem to provide useful hypothetical models of the organization of the universes of the multiverse, as well as of intelligent civilizations in the multiverse and in the spiritual dimensions of the Omniverse. However, sometimes the interdimensional information appears to be mixed with intentional misinformation or disinformation, the intent of which is not entirely clear, although it may be ultimately intended to protect humans from the consequences of having such information.

A good example of such apparently intentional mixing is the *Urantia Book,* a 2,097-page text of interdimensional information published in 1955.[32] On one level the *Urantia Book* provides extremely valuable hints about possible cosmological models of the Omniverse. Yet it maintains that the Omniverse is structured around seven

Superuniverses encompassing planets of the time-space dimensions. Each Superuniverse is composed of about 100,000 local universes for a total of 700,000 local universes, each of which contains approximately 10 million inhabited or inhabitable planets.[33] This is a far smaller number than the one calculated by Professors Linde and Vanchurin.[34]

Would the interdimensional authors of the *Urantia Book* have engaged in intentional deception about the number of universes in the multiverse, and why? Well, one answer is that revelations such as those of the *Urantia Book* are intended to transmit general models of the cosmos as hints to humanity. Perhaps its authors were not permitted to reveal detailed information such as Linde and Vanchurin were to discover. This same pattern has been discovered with Earthling humans who have been contacted by hyperdimensional advanced intelligent civilizations: they have often been given accurate information mixed in with intentional misinformation. One purpose of this strategy may be to protect the contactee from terrestrial military-intelligence authorities, as will be discussed in more depth in chapter 5.

Despite the numerous challenges our map of the Omniverse has been developed in great detail. In the next chapter we present a dimension-based typology of the intelligent civilizations of the multiverse.

2

Typology of Intelligent Civilizations

As mentioned in the preface, when an editor for Oxford University Press contacted me in August 2009, I discovered that Oxford University Press wanted a book exploring how the principles of Earth's international law might apply within our universe as well as in all the universes in the exopolitical dimensions of the Omniverse.

The editor emphasized to me that:

I have known about your work for some time, and I would like to know if you have the time to do [a book] to encompass an application of the principles of international law to extraterrestrial contact. This work would be a scholarly monograph written to draft suggestions for a legal and diplomatic framework between earth representatives from the UN/various governments, and extraterrestrial intelligences. . . . (I personally believe the evidence for their existence is beyond a reasonable doubt) . . . Without getting into a debate on whether such intelligences exist, . . . the book would prepare policy makers, lawyers, law teachers, and diplomats with the necessary background to navigate such contacts.[1]

At the time it appeared that Oxford University Press was preparing its strategic plan to include science-based and policy works for the post-extraterrestrial disclosure environment. Given that I was a founder of the scientific discipline of exopolitics, it made sense that Oxford would approach me.

As conceptual work began for such a legal and diplomatic framework between Earth representatives from various governments and extraterrestrial intelligences, it appeared that no adequate typologies of extraterrestrial intelligences had so far been developed for this task. A new typology of extraterrestrial intelligences needed to be created and developed.

Conventional typology is "the study of types." In anthropology a typology is "division of culture by races"; in archaeology a typology is "the classification of things according to their characteristics"; and in psychology a typology is "the classification of people according to personality types."[2]

Existing typologies of intelligent civilizations are limited and do not reflect the manner in which intelligent civilizations exist and are self-governed in our universe. The intelligent civilizations in our universe self-report themselves as being based in a specific dimension or density of consciousness. That specific dimension or density is that civilization's most fundamental typological, locational, and developmental criterion.

Accordingly, I (with immense help from colleagues) have developed a dimension-based typology of intelligent civilizations, using recent whistle-blower, direct witness, and documentary evidence as sources. This new typology establishes dimension-based criteria for extraterrestrial and interdimensional civilizations and extraterrestrial governance bodies.[3]

Complementary Typologies

Before presenting the dimension-based typology, a brief review of a few previously developed typologies will provide helpful background.

KARDASHEV SCALE

An earlier, complementary typology for mapping intelligent civilizations is the Kardashev scale, first proposed in 1964 by the Soviet Russian astronomer Nikolai Kardashev and popularized by Michio Kaku, Ph.D. It is based on "a civilization's level of technological advancement." The scale is only theoretical and, in terms of an actual civilization, highly speculative; however, it puts energy consumption of an entire civilization in a cosmic perspective. The scale has three designated categories called Type I, II, and III. These are based on the amount of usable energy a civilization has at its disposal and the degree of space colonization. In general terms a Type I civilization has achieved mastery of the resources of its home planet, Type II of its solar system, and Type III of its galaxy.[4]

THE SALLA TYPOLOGY

Another earlier typology classifies intelligent civilizations by race or approximate location in the universe. This typology, developed by Michael E. Salla, Ph.D., also asks whether a specific extraterrestrial civilization is "cooperating with or outside Earth's Military-Industrial-Extraterrestrial Complex."[5] This Earth-centric typology does not include the dimensional anchoring of intelligent civilizations as a fundamental criterion of a civilization's reality. The typology is limited to those civilizations that are in relationship to the governing complex of one nation (USA) on one planet (Earth) in one solar system (Sol) in a galaxy (Milky Way) that has been shown to have at least 500 million Earth-like planets and a universe that may have hundreds of billions of galaxies. This specific typology in its present form is too Earth-centric for application to our universe, let alone the Omniverse.

DENSITY-BASED ECOLOGIES OF CONSCIOUSNESS

The literature on the density-based ecology of consciousness system is congruent with the dimensional ecology of the Omniverse hypothesis. It describes densities or octaves of levels of energies of consciousness, which interact. Philosopher George Gurdjieff (1866–1949) also developed systems for stratification of consciousness that are congruent with the dimensional ecology of the Omniverse hypothesis. The major sources to date of descriptions of the density-based systems of consciousness are channeled or interdimensionally derived information.

For example, cosmologist David Wilcock has constructed a density-based model of our universe, although he does not address the multiple universes of the multiverse or the spiritual dimensions of the Omniverse. His model is based on channeled material (*The Law of One*) and on replicable empirical evidence, including that of Michael Newton, Ph.D. Wilcock writes:

> Although this is not considered a popular notion by any means, there is extensive scientific proof that the universe itself is alive. It is a vast, singular, living being—and we are far more interconnected with it than we may have ever believed. A unified, conscious energy field generates the entire cosmos, and this hidden energy can appropriately be called the Source Field. In this new science, galaxies, stars, and planets are life-forms on a scale we can barely even imagine. The basic laws of quantum physics arrange DNA and biological life out of atoms and molecules we would normally consider to be inanimate.[6]

In the density-based models of consciousness local to our universe, "density denotes a qualitatively distinct level of being. Each density has its own structure of life forms, perception, and typical lessons for the consciousnesses residing in it."[7] The nature of the various density levels of consciousness according to these models can be described as follows:

First density. The first density is the consciousness base of inanimate matter and energy local to our universe in the exopolitical dimensions.[8]

Second density. The second density is:

the base density of the evolution of consciousness in the vegetable and animal kingdoms, for example, local to our Earth and our universe. The evolution of consciousness in the second density concerns biological life, survival, adaptation, competition, group organization as seen with animals, and the like. The soul structure is generally a species soul pool, but as species become more advanced, individual members of the species may differentiate themselves by more varied individual learning.[9]

Third density. The third density:

corresponds to living beings that have a degree of individual consciousness and corresponding free will, at least potentially. This implies that a proper being of 3rd density is a moral entity with attendant responsibilities. While the animal does primarily [act] according to the typical behavior of the species, the 3rd density entity is supposed to have reflexive self-awareness and free will and to be thus accountable in terms of karma and soul evolution.[10]

Fourth density. The fourth density is:

a partly physical state where graduates of 3rd density may deepen and perfect their chosen polarity. Service To Self and Service To Other groups are distinct in 4th density and do not automatically come in contact, unless in the context of interacting with 3rd density. [Fourth-] density beings enjoy more conscious control over physicality and generally form groups telepathically sharing a common pool of experience while retaining a certain individuality. It seems that 3rd-density beings may visit 4th-density conditions in the context of UFO abduction, for example. This does not, however, make one a 4th-density being. The density to which one is native depends on development of consciousness.[11]

Fifth density. The fifth density of the density-based system is termed the interlife (or afterlife) in the dimensional ecology of the Omniverse hypothesis. According to the fifth-density model:

> Souls of 1st thru 4th density find themselves in 5th density between incarnations. This is a contemplation zone where these souls may observe their past/future lives from a purely ethereal state of being. However, for progress to be realized, the souls must incarnate in the density which best corresponds to their level of progress.[12]

Sixth density. According to the density-based system of consciousness, the sixth density is:

> a level of non-material existence where souls have outgrown the need to incarnate in any density. Souls of 1st through 4th densities go through incarnations, with a contemplation period as non-material forms in fifth density in between. 6D is the last stage before union with the One, or seventh density. Seventh density would correspond to the one source of all creation. The archangels or solar world of the Gurdjieffian cosmology may refer to sixth density.[13]

Seventh density. According to the density-based system of consciousness, this is:

> the level where all is one and one is all, in a practical, real and meaningful sense. There is no longer any difference between thought and reality. This corresponds to a notion of all encompassing god or universe or the Sun Absolute of Gurdjieff.[14]

According to one model of consciousness as defined by density, "'density' is more or less the same as 'dimension.' . . . 'Density' denotes a vibrational frequency and not a location, which the term 'dimension' implies. The density structure of this reality is primarily expressed in . . . levels, though each level has sublevels within it. The density scale is a model used to communicate one's perception or orientation in relation to other realities."[15]

The literature on dimensional and density-based ecologies of consciousness can be verified and tested against soul memories of the interlife derived through hypnotic regression. That literature can also serve as a useful guide in suggesting future directions for interlife (afterlife) research based on soul memories of the interlife or other quantitative methods.[16]

Dimension-Based Typology of the Intelligent Civilizations

The dimension-based typology of intelligent civilizations establishes an evidence-based typological model for intelligent civilizations and for intelligent civilizations' governance authorities. This new typology of intelligent civilizations is based on how advanced civilizations describe themselves. The exopolitical literature shows, for example, that advanced intelligent civilizations describe themselves by the dimension in which they are based. For example, "I am a Pleiadian of the fifth dimension" or "I am an Arcturian of the sixth dimension."[17] Our Earth's time-space dimension has been variously described as the third or fourth dimension.*

The new dimension-based typology of intelligent civilizations integrates whistle-blower, direct eyewitness, and documentary evidence for intelligent civilizations in the parallel universes of the multiverse. The new typology establishes an evidence-based typological model for intelligent civilizations. In outline format the dimension-based typology of intelligent civilizations in the exopolitical dimensions of the Omniverse is as follows:

---

*See, for example, Clow and Clow, *Alchemy of Nine Dimensions* and *The Pleiadian Agenda;* Brown, *Cosmic Voyage* and *Cosmic Explorers.*

A. Intelligent civilizations (time-space dimension)

1. Solar-system civilizations based in the time-space dimension, such as:

   a. our human civilization here on Earth

   b. the intelligent human civilization living under the surface of Mars that reportedly enjoys a strategic relationship with the United States government

2. Deep-space civilizations based in the time-space dimension on a planet, solar system, or space station in our or another galaxy or in some other location in this known physical universe

B. Hyperdimensional civilizations: intelligent civilizations that are based in dimensions higher than our time-space dimension or in other parallel universes and that may use technologically advanced transport or physical forms when entering our known time-space dimension

C. Intelligent civilizations' governance authorities: the dimension-based typology identifies legally constituted governance authorities with jurisdiction over a defined dimensional component of our universe or another parallel universe of the multiverse

As verifiable disclosures of classified U.S. and other governmental reports on human-extraterrestrial liaison and planetary visitation programs are made public, directly or through whistle-blower eyewitnesses, the dimension-based typology can be expanded and refined. Likewise, as more data about the nature of intelligent civilizations in universes parallel to our own are gathered, the dimension-based typology of intelligent civilizations in the universes of the multiverse can be refined and expanded accordingly. Replicable, empirical communication with representatives of governance authorities of our Milky Way galaxy has already been achieved using standard laboratory remote-viewing protocols, according to reports in the public domain.[18]

Exophenotypology

Intelligent Civilizations in the Exopolitical Dimensions

The dimension-based typology of intelligent civilizations in the multiverse helps us identify the dimension, density, or strata in which a particular intelligent species or civilization is based. Concurrent with its dimension, an intelligent species or civilization can also be typed by exophenotypology, which is "the typology or classification of extraterrestrials based on their physical appearance," which has been developed by Manuel S. Lamiroy of Exopolitics South Africa.[19] According to Lamiroy, Patrick Huyghe and Dennis Stacy were probably the first to make a systematic classification based on phenotypes, that is, the observable characteristics of aliens.[20]

Higher intelligence and intelligent civilizations can manifest in our universe and universes of the multiverse in many different exophenotypes. A preliminary exophenotypology of intelligent civilizations in the multiverse has been developed, based on observed intelligent civilizations' species.[21]

HUMANOID EXOPHENOTYPE

Humanoid exophenotypes are those who physically resemble our species of human beings. Four subcategories of the humanoid exophenotype can be distinguished, as follows:

> **Standard terran humanoid:** These are the extraterrestrials who look identical to terrestrial human beings and who can, therefore, walk among us unrecognized. This subcategory includes the so-called Blond or Nordic extraterrestrials who are often observed, as well as many others. A humanoid exophenotype

observed on Earth is the Sasquatch, Ogo, or Yeti. These are three names for what is likely just one species, in three different locations, and with a different color of fur depending on the location: white fur in the Himalayas, reddish brown to black fur in North America, and black fur in Africa. People also refer to these beings as "Bigfoot" and, inspired by *Star Wars,* as "Chewbaccas" or "Wookiees."[22]

Shorter humanoid: Terrestrial-looking extraterrestrials that are shorter than most terrestrial humans. These include, for example, the Santinian, the Korendrian, and the Short non-Grey.

Taller humanoid: This subcategory consists of extraterrestrials that look terrestrial in appearance, but are taller than the average human being. These include the Tall White, as well as several types of "giants" (e.g., the Bawwi).

Humanoid with different features: A fourth subcategory consists of beings that in many ways look like terrestrial humans, yet have one or more clearly distinguishing features, such as colors or types of skin, hair, or eyes that are not found here on Earth. These include, for example, cat people, bird-people, blue-skinned aliens, and aliens with wrinkled faces.[23]

Five Martian humanoid exophenotypes, discovered by former U.S. chrononaut Andrew D. Basiago and discussed at length in the next chapter, also fall under the humanoid exophenotypes:

> ✸ *Homo martis terries*
> ✸ *Homo martis martis*
> ✸ *Homo martis extraterrestrialis*
> ✸ *Homo martis deminimis*
> ✸ *Homo martis sasquatchis*[24]

GREY HYPERDIMENSIONAL EXOPHENOTYPE

According to the exophenotypology developed by Lamiroy,

> The Greys are probably the best-known phenotype of extraterrestrials
> [or hyperdimensionals]. Older UFO literature distinguishes three main
> types, but closer inspection reveals there are actually five, often with
> many variations between them, especially for the "standard" Greys,
> who differ in skin color, in whether their arms seem to have an elbow
> or not, in whether they have hands or claws, and in the appendages on
> those hands, which can be fingers or tentacles with suction caps.[25]

The five main types are as follows:

Standard Greys are the most common type. They are about 4 to
4.5 feet tall, with large bulbous heads and wraparound eyes, a
slitlike mouth, small ears without earlobes, and no visible nose.
Their legs are shorter and jointed differently than one would
expect in a human. Their arms often reach down to their knees.
These Greys have been seen in many different colors and are
generally pale in appearance: gray, white, (pale) blue, (pale)
green, (pale) orange, and brown. There does not seem to be one
standard phenotype of these, as they have been observed with
many different features: some have hands with fingers, some
have claws, some have webs, and so on.

Tall Nosed Greys are 6 to 9 feet tall and, unlike other Greys, have
a distinct nose.

Tall Greys are 6 to 7 feet tall and basically are taller versions of
the Standard Greys. They have large heads, wraparound eyes,
and a slitlike mouth. There may actually be two different types
because, unlike the Standard Greys, some of these Tall Greys
are said to have reproductive organs. They also seem to have
something that resembles a small nose.

Short Greys are shorter, about 3.5 feet tall. They typically look like shorter, but far more muscular, versions of the Standard Greys. They are said to be extremely aggressive and generally are believed to be the most dangerous of all Grey species. (On the Internet these are often dubbed "Bellatrix Greys.")

Mini Greys: in recent years, there have been reports of Greys that fit the description of the Standard Greys, but are only two feet tall instead of the usual 4 to 4.5.[26]

Although some entities known as Grey hyperdimensionals may be artificial cyborgs or probes, the Grey races, which one expert numbers in excess of 150, are a living species. A dialogue with a Short Grey sponsored in 2006 by the Exobiology Project—a project involving volunteer researchers into exophenotypes and exobiology and including former U.S. Air Force officers—suggested that Greys, while not plants, have a biology that thrives on chlorophyll, as this brief excerpt shows.

Q: Are you a mechanical species, as some Humans claim?

SMALL GREYS: No, we are biological but different in our biology than Human and other species.

Q: Are you part plant, as some humans say, that your tissues contain chlorophyll?

SMALL GREYS: We are not plant but use chlorophyll in a symbiotic way. Chlorophyll provides us with energy and nutrients.

Q: Is that why you look bluish-grey in white light?

SMALL GREYS: Yes, the chlorophyll in our skins colors our appearance. We are also able to bioluminescence.

Q: I once "observed" a Grey in captivity. Did this one die and what could have been done to keep it alive?

SMALL GREYS: It did die. It needed other Greys and it needed nutrients.

Q: Do we have these nutrients on Earth?

SMALL GREYS: Some of them. Chlorophyll, boron, but others you do not have.

Q: So, that means that captured Greys will always die?

SMALL GREYS: Not always. Some may have enough reserves to survive.

Q: Are you mainly nocturnal?

SMALL GREYS: Yes, we work by night. Bright light, unless it has special filters, damages us. Both heat and light damage us.

Q: How does the chlorophyll provide nutrients if it does not have sunlight?

SMALL GREYS: We have special illumination.[27]

REPTILIAN EXOPHENOTYPE

Current literature about alien abduction and encounters mentions beings that are reptilian in appearance, which are called reptilian, reptoid, dinoid, or saurian. Virtually all of those mentioned below are bipedal—they walk upright—and therefore are not really reptiles, but instead have characteristics reminiscent of reptilians. The main two exceptions to this would be the Ciakar and the "giant snakes," both of whom fit the description of actual reptiles.

According to the exophenotypology developed by Lamiroy, the following subcategories of the reptilian exophenotype can be distinguished.

Tall Winged Draco (Ciakar). These have been mentioned in the literature, but have not actually been encountered on Earth. They look like giant lizards, range from 14 to 22 feet tall, can weigh up to 1,800 pounds, and can have winged appendages.

Winged Draco/Mothmen. Reports about flying reptilians with wings originate mainly in the Americas and in Asia. These reptilians are bipedal and have a very muscular, athletic build. They stand 7 to 8 feet tall. Their heads are typically said to be dinosaur-like (like *Tyrannosaurus rex*). One feature mentioned again and again is that they have glowing red eyes with a hypnotic stare. Like small Greys they are said to be able to fly or levitate. They have wings on their backs, but thus far there have been no reports of these wings moving or being used, not even when the beings are flying. The literature often refers to these beings as Mothmen.

Non-winged Draco. The non-winged Draco looks much like the winged Draco without wings and different-colored eyes. The non-winged Draco are one of the reptilian species most often seen on board ships that abductees are taken to. As such, they are often seen in the company of Greys. Reports of these non-winged Draco come from all over the world. They are among the most commonly encountered reptilians. The reports are contradictory, however, over whether these reptilians have genitalia. Some accounts mention no genitalia, while others do mention them having genitalia and engaging in sexual acts with abductees.

Iguana-like. The literature also mentions reptilians that are described as iguana-like. They look similar to the non-winged Draco, but are less muscular. Their heads resemble that of an iguana more than that of a Tyrannosaurus. They are also shorter than the typical non-winged Draco, being about 5 feet tall. In most reports they are said to be hooded.

Lizardlike. A fifth subcategory of reptilians consists of lizard men/ snake men. They are about 6 to 8 feet tall, upright, bipedal, often have lizardlike scales, are greenish to brownish in color, and have clawlike, four-fingered, webbed hands. Their faces are said to be a cross between that of a human and that of a snake or lizard. Some are said to have a central ridge coming down from the top of their head to their snout. Since some witnesses refer to a "snakelike" head, whereas others use "lizardlike," these may be two separate groups. Given the different descriptions of skin and genitalia, there actually may be several different groups.

Giant snakes. Some unconfirmed reports on the Internet mention gigantic snakes of extraterrestrial origin. These would be genuine reptiles, not bipedal, not even legged.

Others. Abductees have also described encounters with reptilian-looking beings that do not fall in the categories listed above. Some of them seem to be variations on the other phenotypes. For example, there are accounts of beings who fit the description of the lizard men, but who seem to have a wet leathery skin and hands with fingers instead of claws.[28]

A 2006 dialogue conducted by the Exobiology Project with a reptilian hyperdimensional set out several possible characteristics of this species.

Q: You sound mechanical, according to Earth characteristics. Are you mechanical or physical?

HOODED REPTILIAN: We are physical. We are not mechanical.

Q: Why are some species interested in hydrocarbons? Are they rare in the galaxy? What about other minerals and gases: hydrogen, boron? What purpose do these species put hydrocarbons to? Are we in conflict for these resources?

HOODED REPTILIAN: The Greys bring Earth matters to us for

fuel, for nurture, for cleansing, for breathing, for sustenance, for "greening."

Q: What is "greening"?

Hooded Reptilian: We have chlorophyll in our skins for nutrition.

Q: Like the Greys?

Hooded Reptilian: Yes. There is no conflict for resources. Resources are everywhere.

Q: That is what the other races have said.

Hooded Reptilian: That is true.[29]

The Lamiroy exophenotypology also develops insectoid and hybrid typologies.

Insectoid Exophenotype

The insectoid phenotype consists of those beings whose appearances (and/or behavior) resemble those of insects. As with the reptilians and amphibians, the insectoids are a mixed category, in which some phenotypes display some humanoid traits, whereas others are more akin to what is normally considered typical for an insect.

Several types have thus far been included:

Praying mantis–like. The most common insectoids are beings who are reminiscent of praying mantises. They are encountered quite often during abduction experiences. Most accounts describe them as approximately 6 feet tall. Most reports list them as bipedal, though some others describe them as more insectlike. It is possible that the bipedal ones could be hybrids.

Depanoid. There are reports (and photographs) of beings that are bipedal, with a humanoid torso and limbs, but with an oversized "bug" head. These, too, have been encountered by a handful of abductees. The name "depanoid" was given by one of those abductees, Jim R. The beings are described as "energy-suckers" or "energy vampires," who put their victims through experiences that the victims respond highly emotionally to; then the beings "feed" off those emotions.[30]

HYBRID EXOPHENOTYPE

According to Lamiroy, "A hybrid is anything derived from heterogeneous sources or composed of elements of different or incongruous kinds. The hybrids spoken of in UFO literature are primarily a cross between Earth human and Zeta Reticuli beings. The specific process that is used to create these hybrids has not yet been revealed. It uses not only genetic splicing and cloning, but a form of light-plasma engineering technology with which humans are unfamiliar."[31]

OTHER

There also exist exophenotypes for cetaceans (whales and dolphins), amphibians, robots, clones, orbs, shadow beings, stick figures, and nonphysical beings and groups.

Earthling Human Sightings of and Contact with Varied Exophenotypes of Intelligent Civilizations

As will be discussed more fully in chapter 4, there is a broad literature documenting Earthling human sightings of and contact with a variety of exophenotypes of intelligent civilizations.[32] A standard research database on sightings of possible intelligent civilizations from the dimensional ecology states: "The number of documented general sightings in the modern age numbers approximately 120,000, of which 20,000 have been described as landings. Numerous surveys and opinion polls conducted over the last 50 years consistently report that between [5 to 10 percent] of the US population has seen what they consider to be a UFO. According to the United Nations, since 1947, approximately 150 million people have witnessed UFO sightings throughout the world."[33]

About Earthling human sightings of other humanoid exophenotypes, the research database states, "There are at least 4,000 humanoid cases (researched by Albert Rosales), also known as Close Encounters of the Third Kind (CE-III), a term coined by J. Allen Hynek. 'Humanoid' means 'having human form or characteristics.' The term 'Close Encounter of the Third Kind' means seeing or coming in contact with an entity of unknown origin. The encounter may or may not be related to the sighting of an unidentified craft."[34]

For sightings by qualified pilots, the research database states:

There have been over 3,500 documented sightings of Unidentified Aerial Phenomena by military, civilian, and commercial airline pilots. These observations span the entire history of powered flight. Many of these cases come from declassified US government reports and investigations, international reports from official sources, and the direct testimony of military and commercial pilots, air traffic controllers, and radar operators.[35]

On Earthling human contacts with a diversity of exophenotypes, Lamiroy recently wrote, "Work on the typology has been partially suspended since March 2012 after coming across a database of more than 17,000 contact cases, which literally includes dozens of phenotypes that will still have to be integrated in the exophenotypology."[36]

From the platforms of the dimensions in which they are based, various hyperdimensional civilizations reportedly carry out "interdimensional teleportation" on large numbers and categories of Earthling humans.[37] There is replicable prima facie empirical evidence to show that a low estimate of 2 percent of the U.S. population and a high estimate of 14.7 percent of the total human population may have had their physical bodies, etheric bodies, or consciousness temporarily or permanently teleported to spaceships or dimensions other than our time-space Earth dimension by hyperdimensional civilizations.[38] The eyewitnesses to these teleportations reply that they are carried out by various species of intelligent hyperdimensional civilizations of the Grey exophenotype, as well as by hyperdimensional civilizations of other exophenotypes. Expert estimates of the number of known Grey intelligent species range as high as 150 different species.[39]

PART 2

Exopolitical Dimensions of the Omniverse

3

Time-Space
Solar-System Civilizations

Our initial focus in the dimension-based typology of intelligent civilizations in the Omniverse is the civilizations based in the time-space dimension in our solar system, Sol. There is newly emerging presumptive empirical evidence that time-space-dimension intelligent indigenous human and humanoid civilizations exist on Earth's nearest planetary neighbor, Mars. The existence of our human civilization on Earth can be demonstrated by self-evident common sense, empirical science, and the law of evidence. For example, the science of anthropology confirms the existence of *Homo sapiens* on Earth.[1] In the same way the empirical science of exopolitics and the law of evidence can provide persuasive evidence for the existence of intelligent life on Mars.[2]

Proof of the Existence of
Intelligent Human Martian Life

Just as skeptics state that the available evidence proves there is no current indigenous intelligent life on Mars (other than the secret U.S. space program), we can present evidence for the opposite of this assertion and thus prove that there is indigenous intelligent life of the humanoid exophenotype currently on Mars. All sides of the debate about the existence of such life deserve to be respectfully heard according to the methodology of the law of evidence.

New scientific data and recent witness and documentary evidence establish that an indigenous human civilization, as well as indigenous, intelligent extraterrestrial life, both exist on Mars, based in cities beneath the surface, along with humanoid species on the surface. This new generation of data supporting the existence of Martian life has been emerging first from independent whistle-blowers who were involved in compartmentalized secret U.S. government Mars programs and second from the NASA Mars rovers, Spirit, Opportunity, and Curiosity, which have been on the Martian surface, photographing it at close range.

The first formal exophenotypology of intelligent humanoid beings on Mars is based on eyewitness evidence from the two early secret U.S. government quantum access time-travel and teleportation programs, both DARPA's Project Pegasus and the CIA jump room program. The CIA's Mars program consisted of "jump room" teleportation that had reportedly been designed and given to the U.S. government by a specific species of the Grey exophenotype hyperdimensionals.* It included jump room facilities in New York City and Los Angeles, California, and a mission-control facility in Ohio.[3]

At this historical stage in the exploration of Mars, the appropriate standard of proof in determining whether the available witness and documentary evidence shows that indigenous intelligent life does exist on Mars is persuasive evidence on the face of it. There is no scientific, legal, or ethical requirement that such a determination must satisfy the whimsical criterion proposed by Carl Sagan that "Extraordinary claims require extraordinary evidence."[4]

---

*The evidence for hyperdimensional civilizations and their interactions with time-space intelligent civilizations in our universe is presented in chapters 4 and 8.

Eyewitness Evidence of
Martian Humanoids

Andrew D. Basiago is the first Earth person to summarize publicly the five exophenotypes of bipedal humanoids on Mars. He is a Cambridge-trained environmentalist and attorney at law (admitted to the Washington State Bar and the United States District Court for the Western District of Washington). He was one of America's early time-space explorers and the first American child to teleport. Mr. Basiago served in DARPA's Project Pegasus from 1968 to 1972, at ages 7 to 11; he was involved in eight different types of time travel that had only recently been achieved by the U.S. defense-technical community. He also served in a second secret U.S. space project, the CIA's Mars jump room program, from 1980 to 1984, during his college years at UCLA. In that capacity he made about forty trips to Mars via jump room as a U.S. chrononaut.[5]

Images of Martian life-forms, including each of the three exophenotypes of Martian humanoids, were first publicly disclosed by Basiago in his "The Discovery of Life on Mars."[6] It accurately portrayed the Martian humanoid exophenotypes, as later confirmed by lifelong CIA employee Virginia Olds.[7]

As he describes:

There are five species of bipedal humanoid on Mars. They are: (1) The descendants of Earthlings who went to Mars during the first civilization on Earth that were separated from us by the solar system catastrophe around 11,500 BCE [*Homo martis terris*]; (2) A species of indigenous Martians [*Homo martis martis*]; and (3) A stealthy Martian humanoid species that is a branch of the Grey ET's [*Homo martis extraterrestrialis*]. There are two other primates on Mars, too. They are: (4) A tiny species of hominoid that burrows in the ground [*Homo martis deminimis*]; and (5) A Bigfoot-like primate

[*Homo martis sasquatchis*]. Mars is their planet. It doesn't belong to us. Mars belongs to the Martians.[8]

Basiago developed his exophenotypology of Martian humanoids from direct eyewitness observation of the *Homo martis terris* exophenotype on Earth during his service as a participant on Project Pegasus, as well as from direct observation of the *Homo martis martis* and *homo martis extraterrestrialis* types on jump room trips to Mars. Basiago also benefited from other accounts of Martian humanoid exophenotypes, including official briefings by the CIA in preparation for Mars visits, and informal briefings with CIA officers, jump room project personnel, and participants both on Earth and on Mars.

Basiago has publicly confirmed that in 1970, in the company of his late father, Raymond F. Basiago, an engineer for the Ralph M. Parsons Company who worked on classified aerospace projects, he met three astronauts of the Martian human civilization at the Curtiss-Wright Aeronautical Company facility in Wood Ridge, New Jersey, while the Martians were there on a liaison mission to Earth and meeting with U.S. defense-technical personnel. These astronauts were of the *Homo martis terris* exophenotype. He describes them:

> *Homo martis terris* are our genetic relatives from the time before the solar system catastrophe of 9,500 BC, when Earth and Mars were in contact. They resemble bald, homely Earthlings. . . . They are human beings, like human beings from Earth. It was clear that by 1970 they both had a more advanced space-faring capability than we did on Earth and were working with the United States government.[9]

Basiago reports observing two of the three principal exophenotypes of Martian humanoids while on Mars during trips in service on the CIA Mars jump room program. These two are *Homo martis martis* and *Homo martis extraterrestrialis:*

> *Homo martis martis* are the indigenous Martian humanoids. They have narrower heads, pointier ears, longer fingers, and smaller bodies than Earthlings. They resemble the creature in the vintage film *Nosferatu*. We were advised that if hungry enough, they would sometimes kill and eat human visitors from Earth, so we should be wary of them. . . . They gave rise to the notion of "Martian" in folklore and popular culture on Earth. . . .
>
> *Homo martis extraterrestrialis* resemble the Grey ETs of the UFO contact literature. This is the type that Courtney M. Hunt and William B. Stillings and I of the CIA's jump room program encountered after I saw one that was sitting on the roof of the jump room facility called The Corkscrew when we arrived on Mars one time in 1981–82. As I exited The Corkscrew ahead of my fellow chrononauts, I called out to them, "Court, Brett! A Grey. On the roof. Observing us!" They are probably the result of a branch of the Greys that was left back on Mars at some time in Martian history. It was this type of Martian that is depicted in my historic image, "The Humanoid Being on Tsiolkovski Ridge," which I published in 2008 in my landmark paper "The Discovery of Life on Mars," which is the first image of a humanoid being on another planet ever published on Earth.[10]

Basiago also notes:

> There is a fourth typology of humanoid beings on Mars, the tiny ones that Dave Beamer has detected being crushed under the wheels of NASA's Mars Exploration rovers. I don't include this type of Martian humanoid in this taxonomy because they are far smaller than the three other typologies of Martians and us.[11]

INTERVIEW WITH ANDREW BASIAGO

In an interview published in September 2013, in *Fear and Loathing Long Beach* online magazine, Basiago elaborated on his Mars experiences to interviewer Mike Spent.

Q [MIKE SPENT]: What did you see while on Mars? What is Mars like? Is it inhabited?

ANDREW BASIAGO: Mars has a desert environment with scant water, sparse vegetation, and about 30 major land species. It has a blue sky and soils that range from light pink to rust red. It has three major humanoid typologies, namely, *Homo martis terris,* who are the descendants of the Earthlings who were left on Mars after the solar system catastrophe of 11,500 years ago [9500 BCE]; *Homo martis martis,* who are the indigenous Martians that have bulbous heads, long faces, pointy ears, long fingers, spindly bodies, and fearful personalities; and *Homo martis extraterrestrialis,* who are a branch of the Grey ETs that are adapted to terrestrial life on Mars. During my 40 or so jumps, I encountered several animal species, including a predator that had a head like a *Tyrannosaurus rex* and a supple body, and a plesiosaur that was also a predator, with teeth running all the way down its throat. I met one of the *Homo martis martis* type of humanoid when he met Courtney M. Hunt and I at the jump room facility called the Corkscrew and took us on a tour of his underground civilization, and I would see both adult and children of this type when I would walk through a dilapidated brick city to deliver a data disk to a telecommunications post that we had established there. Once, I saw one of the *Homo martis extraterrestrialis* species when I was exiting the Corkscrew with Courtney M. Hunt and William B. Stillings soon after arriving on Mars via jump room. It was lurking atop the jump room facility.[12]

Q: Have you been contacted by the U.S. government lately about your disclosing these secrets?

ANDREW BASIAGO: Yes. After I published my landmark paper "The Discovery of Life on Mars" in 2008, I was contacted by a senior case officer of the CIA, who confirmed that my paper was highly accurate about life-forms on Mars.[13]

EVALUATING THE WITNESS EVIDENCE FOR INDIGENOUS MARTIAN INTELLIGENT LIFE

The eyewitness evidence of Andrew D. Basiago about indigenous Martian intelligent life, both on the surface and under the surface of Mars, is highly credible. I have had the opportunity to interview, question, and cross-examine Mr. Basiago closely about all aspects of his accounts regarding Project Pegasus and Mars since the year 2000. His verbal and written accounts have been consistent and verified by available witnesses and by documentary evidence, including written versions of his evidence that he has made available to me.

Basiago is extremely conscientious in the execution of his duties as an officer of the court and a member of the bar of the State of Washington. He consistently tells the truth—even punctiliously so—and is a conscientious reporter of his activities. The professional penalties for an officer of the court who engages in public deception or fraud regarding an issue of public importance, such as personal knowledge of indigenous Martian intelligent life, could be grave.[14]

Basiago's participation in time travel and teleportation in Project Pegasus, 1968–72, is corroborated by witnesses. U.S. Army Captain Ernest Garcia, whose career in U.S. intelligence included serving both as a guard on the Dead Sea Scroll expeditions of Israeli archaeologist Yigal Yadin and as the Army security attaché to DARPA's Project Pegasus, confirmed Basiago's participation as a U.S. chrononaut in Project Pegasus. Jean Maria Arrigo, Ph.D., an ethicist who works closely with U.S. military and intelligence agencies, has also confirmed Basiago's participation in Project Pegasus.[15]

Two former participants in the jump room program, former U.S. chrononaut William B. Stillings and former presidential adviser Bernard Mendez, have provided independent confirmation of Basiago's participation in the CIA Mars jump room program. Each of these two credible eyewitnesses to intelligent life on Mars have made their whistle-blower testimonies public at considerable personal risk. The testimonies of these three men about their roles as fellow participants in the 1980s CIA jump room program are mutually consistent and congruent.[16]

History of Exposure of
CIA Mars Jump Room Program

The investigative process by which the secret CIA Mars jump room program resulted in the development of a Martian exophenotypology and became public knowledge is worth examining in its historic context, especially since the process of investigating the CIA Mars jump room program occurred during the administration of a sitting U.S. president, Barack H. Obama, who was himself, under his legal name Barry Soetoro, a CIA Mars jump room participant in the early 1980s.[17]

The CIA jump room started to become public knowledge when the two former participants mentioned above, Andrew D. Basiago and fellow chrononaut William B. Stillings, who was tapped by the Mars program for his technical genius, publicly confirmed that U.S. President Barack H. Obama was enrolled in their Mars training class in 1980 and was among the young Americans from the program whom they later encountered on the Martian surface after reaching Mars via the jump room, during visits from 1981 to 1983. Their astonishing revelations provide a new dimension to the controversy surrounding President Obama's background and pose the possibility that it is an elaborate ruse to conceal Obama's participation as a young man in the U.S. secret space program.

Mars Training Class, Summer 1980

According to Basiago and Stillings, in the summer of 1980 they attended a three-week factual seminar about Mars to prepare them for trips that were then later taken to Mars via teleportation. Remote-viewing pioneer Major Ed Dames, who was then serving as a scientific and technical intelligence officer for the U.S. Army, taught the course. It was held at the College of the Siskiyous, a small college near Mt. Shasta in California. They state that ten teenagers were enrolled in the Mars training program. In addition to themselves, two of the eight other teenagers in Major Dames's class that they can identify today were Barack Obama, who was then using the name "Barry Soetoro," and Regina Dugan, whom Mr. Obama appointed as the nineteenth director and first female director of the Defense Advanced Research Projects Agency (DARPA) in 2009.

As many as seven parents of the ten students, all with ties to the CIA, audited the class. They included: Raymond F. Basiago, an engineer for the Ralph M. Parsons Company who was the chief technical liaison between Parsons and the CIA on Tesla-based teleportation; Thomas Stillings, an operations analyst for the Lockheed Corporation who had served with the Office of Naval Intelligence; and Mr. Obama's mother, Stanley Ann Dunham, who carried out assignments for the CIA in Kenya and Indonesia. From 1981 to 1983 the young attendees then went on to teleport to Mars via a jump room located in a building occupied by Hughes Aircraft at 999 N. Sepulveda Boulevard in El Segundo, California, adjacent to the Los Angeles International Airport.

Obama Identified on at Least Two Visits to Mars

On August 21, 2011, Basiago stated, "Something highly significant has happened, and that is that two individuals from the same Mars training class in 1980 (Basiago and Stillings) have met and are comparing experiences and are able to corroborate not only that they were on the surface of Mars together but that before reaching Mars via jump room they were trained with a group of teenagers that included the current president of the United States (Obama) and director of DARPA (Dugan)."[18]

Stillings's statement, released at the same time, read, "I can confirm that Andrew D. Basiago and Barack Obama (then using the name 'Barry Soetoro') were in my Mars training course in Summer 1980 and that during the time period 1981 to 1983, I encountered Andy, Courtney M. Hunt of the CIA, and other Americans on the surface of Mars after reaching Mars via the jump room in El Segundo, California."[19]

It is not known whether NASA's Jet Propulsion Laboratory, which is located in Pasadena, California, had a hand in selecting the young people for their dangerous interplanetary mission to Mars, but it is conspicuous that all four had Pasadena connections. Basiago's father worked for the Ralph M. Parsons Company, headquartered in Pasadena. Stillings was residing in La Canada, a suburb of Pasadena. Mr. Obama had just completed a year of undergraduate studies at Occidental College in Eagle Rock, near Pasadena. Ms. Dugan was attending the California Institute of Technology, located in Pasadena.

In a statement made Sept 20, 2011, Basiago again confirmed Mr. Obama's co-participation in the 1980 Mars training class, stating,

Barry Soetoro, a student at Occidental College, was in my Mars training class under Major Ed Dames at the College of the Siskiyous in Weed, California, in 1980. That fact has been corroborated by one of my other classmates, Brett Stillings. Two years later, when he was taller, thinner, more mature, a better listener, using the name "Barack Obama," and attending a different college, Columbia University, we crossed paths again in Los Angeles and I didn't recognize him as the person that I had been trained with in the Mars program and encountered on the surface of Mars. In fact, doing so would have been virtually impossible in any case, because measures had been taken to block our later memories of Mars shortly after we completed our training in 1980.[20]

Basiago also states that during one of his trips to Mars via jump room that took place from 1981 to 1983, he was sitting on a wall beneath an arching roof that covered one of the jump room facilities as he watched Mr. Obama walk back to the jump room from across the Martian terrain. When Mr. Obama walked past him and Mr. Basiago acknowledged him, Mr. Obama stated, with some sense of fatalism: "Now we're here!"

Stillings states that during one of his visits to Mars, he walked out of the jump room facility and encountered Mr. Obama standing beside the facility by himself staring vacantly into a ravine located adjacent to the facility.

Basiago thinks that it is virtually certain that Ms. Dugan also went to Mars, because he once encountered her at the building in El Segundo where the jump room to Mars was located, as he was entering the building to jump to Mars and she was exiting it. "I know you!" she said, greeting him as she passed him in the lobby of the building.

In his published interview conducted by Mike Spent mentioned earlier, Basiago gave a more comprehensive list of Mars jump room program participants:

Q [MIKE SPENT]: Who else was involved in the trips? Can you name them all?

ANDREW D. BASIAGO: I can't name them all. I have, however, identified 17 Americans associated with the CIA's Mars jump room program. Those who are known to have taken the jump room to Mars are:

Andrew D. Basiago, former U.S. Chrononaut

Raymond F. Basiago, Ralph M. Parsons Company

Major Ed Dames, U.S. Army

Regina E. Dugan, future director of DARPA

Mary Jean Eisenhower, future People to People International, granddaughter, Dwight D. Eisenhower

Courtney M. Hunt, CIA officer

Linda Hunt

William C. McCool, future U.S. Shuttle Columbia astronaut

Bernard Mendez, U.S. Presidential special assistant

Barack H. Obama, future U.S. President

William B. Stillings, former U.S. chrononaut

Stansfield T. Turner, CIA Director

Michael C. Relfe, U.S. armed forces security on Mars

Arthur Neumann, physicist, Lawrence Livermore Laboratory [teleported to Mars in Lawrence Livermore Laboratory project]

Those who participated but did not take the jump room to Mars are:

Stanley A. Dunham, mother of Barack Obama

Thomas J. Stillings, father of William B. Stillings

Michael Strickland[21]

STRANGERS IN A STRANGE LAND

Basiago, Obama, Stillings, and Dugan went to Mars at a time the U.S. facilities on Mars were rudimentary and resembled the construction phase of a rural mining project. While there was some infrastructure supporting the jump rooms on Mars, there were no baselike buildings like the U.S. base on Mars first revealed publicly by Command Sgt. Major Robert Dean at the European Exopolitics Summit in Barcelona, Spain, in 2009. The primitive conditions that they encountered on

Mars might explain the high level of danger involved. Basiago and Stillings agree that Major Dames stated during their training class, "Of the 97,000 individuals that we have thus far sent to Mars, only 7,000 have survived there after five years."[22]

In light of these risks, prior to going to Mars, Basiago received additional training from Mr. Hunt, a career CIA officer, who showed Basiago how to operate the respiration device that he would wear only during his first jump to Mars in July 1981, provided him with a weapon to protect himself on Mars, and took him to the Lockheed facility in Burbank, California, for training in avoiding predators on the Martian surface.

When they first teleported to Mars in summer 1981, the young Mars visitors confronted the situation that Major Dames had covered at length during the class the previous summer, that one of their principal concerns on Mars would be to avoid being devoured by one of the predator species on the Martian surface, some of which they would be able to evade, and some of which were impossible to evade if encountered.

The Mars program was launched, Basiago and Stillings were told, to establish a defense regime protecting the Earth from threats from space and, by sending civilians, to establish a legal basis for the United States to assert a claim of territorial sovereignty over Mars. In furtherance of these goals and the expectation that human beings from Earth would begin visiting Mars in greater numbers, their mission was to acclimate Martian humanoids and animals to their presence or, as Major Dames stated during their training, "Simply put, your task is to be seen and not eaten."[23]

DENIALS AND CONFIRMATIONS

On November 8, 2011, I was the first to report revelations made by Basiago and Stillings that Obama had served with them in the jump room program of the early 1980s. The two chrononauts then appeared on late-night talk radio's Coast-to-Coast AM with Laura Magdalene Eisenhower, the great-granddaughter of U.S. President Dwight D. Eisenhower, to discuss their involvement in the U.S. secret space program.

Major Dames called the show to deny his involvement in the jump room program. Then, on January 3, 2012, Tommy Vietor, the official spokesperson of the U.S. National Security Council—a body that includes the U.S. president, vice president, secretary of defense, secretary of state, and national security adviser—denied that Obama had participated in the CIA Mars jump room program. With multiple whistle-blowers coming forward and corroborating each other's testimony, it now seems inevitable that both the cover-up of the U.S. presence on Mars and Mr. Obama's personal involvement in it will soon become matters of great public interest.

According to Huffington Post UK reporter Michael Rundle:

The White House has denied that President Barack Obama was teleported to Mars in the 1980s as a member of a secret CIA project hosted at a community college in California. Two people have made the claim while asserting that they themselves were time-traveling government agents sent across the solar system to explore new worlds. Andrew D. Basiago and William Stillings have argued that they served as "chrononauts" at the request of the Defense Advanced Research Projects Agency. The two men also insist that the Commander in Chief was part of the team and made two trips to the Red Planet via a "jump room." At that time, aged 19, Barack was reportedly known as Barry Soetoro. . . .

Now *Wired* magazine reports that the White House has been forced—albeit with tongue firmly in cheek—to deny that Obama ever made the 200 [million] mile trip. So has Obama been to Mars? "Only if you count watching Marvin the Martian," Tommy Vietor, the spokesman for the National Security Council, told *Wired*.[24]

Following the January 2012 White House denial, the next phase of the CIA's Mars jump room and Barack Obama's participation in it becoming public knowledge began with the appearance of Bernard Mendez to corroborate the existence of the program and Basiago's accounts of it.[25] He made these allegations during a seminar held in Vancouver, British Columbia, on June 1, 2012, under the auspices of the ExoUniversity.org. According to Bernard Mendez's biography:

Bernard Mendez was born into a family involved deeply in military service. His father, Bernard Mendez, Sr., served in World War II for the US Army and was intimately involved in secret military projects.

At the age of two, Mendez was observed by family members to have a seeming ability to see objects in the invisible range of the spectrum. Upon medical examination, it was discovered that he possessed an unusually high number of rods and cones in his eyes that allowed him to see in the dark.

Throughout his youth, Mendez had numerous direct encounters with several ET races, most of which were Greys. This continued into his teenage years.

By the age of 16, Mendez had enlisted into the United States Navy, where he was assigned to the Special Forces command. On October 9, 1970, he was part of a very bizarre abduction incident in Australia and [was] eventually rescued by Navy Seals, as reported in the Associated Press. This incident caught the attention of then-President Richard Nixon, who asked to meet the young sailor at the White House.

Eventually, Nixon recruited Mendez to be a special assistant in a secret government program dealing directly with extraterrestrials. He was sent to locations all over the world, including Alaska, Australia, China, France, Antarctica, the North Pole, Catalina Island, K2, the Fjords, as well as the infamous secret underground bases known as Area 51 and Camp Hero (aka Montauk Naval Air Station), Groom

Lake, China Lake and Clove Lakes. He visited these places in order to negotiate directly with various ET representatives on behalf of the US government. Mendez even visited a secret underground base in England known as the "Bat Cave" and the former Soviet Union's version of NASA, Kapustin Yar, near Volgograd, Russia.

During his years of service, Mendez obtained what is known as a cosmic-level clearance, designated "Yankee White," which permitted him to work for the White House. He rose through many different access levels . . . before eventually reaching Access Level 59. He was told he would go no further even though there was an Access Level 60, which was the highest level.

From the spring of 1971 until the Watergate scandal forced President Nixon to resign on August 9, 1974, Mendez worked for the Nixon White House as a go-between with various Greys and Pleiadians. He continued in secret government service for a brief period of time into the 1980's before retiring at a young age.

In 2011, Mendez heard an interview of Andrew D. Basiago on talk radio's Coast to Coast AM with George Noory about his participation in a secret government time travel program called Project Pegasus in the early 1970's and a later secret Mars visitation program in the early 1980's. Mendez contacted the American chrononaut and they confirmed that they had served together in Project Mars, which was the CIA's Mars jump room program of the early 1980's, with Barack H. Obama, Regina E. Dugan, William B. Stillings, William C. McCool, and other young Americans of their generation.[26]

During the June 2012 seminar, Mendez confirmed Basiago's and Stillings's testimony, stating that he, Mr. Obama, former DARPA director Regina E. Dugan, Andrew D. Basiago, and William B. Stillings attended the jump room training class taught by Maj. Ed Dames at College of the Siskiyous in the summer of 1980. He also stated that he teleported with Mr. Obama and Mr. Basiago in jumps that took place from 1981 to 1983, when Obama and Basiago were college students. At the time Mendez was investigating the jump room program for the U.S. intelligence community.

Mendez's statements confirming the existence of the CIA's jump room program and the roles played in it by Major Dames and Mr. Obama directly refute their denials. Other jump room program participants identified by Mendez and Basiago have neither confirmed nor denied their involvement in the program as of this writing.

Asked for comment on Obama's silence in the face of confirmation of his participation by three jump room participants, Basiago stated, "President Obama campaigned for president pledging that his administration would be one of transparency. He must now fulfill that promise and acknowledge that he served with us in the jump room program."[27]

MENDEZ'S SPECIAL JUMP ROOM MISSION

During the ExoUniversity.org seminar, Mendez explained that his mission to evaluate the CIA's jump room program originated from discrepancies being reported by the jump rooms on the U.S. East and West Coasts. He said that the East Coast jump room was located in New York City, the West Coast jump room in El Segundo, CA.

According to Mendez the jump room technology had been transferred from a species of Grey extraterrestrials to the U.S. government. The West Coast and East Coast jump rooms were reporting up to forty incidents of participant injuries per month occurring during jump room "teleportation" to unknown environments in space. Yet, several days after each teleportation jump, the reported injuries to participants had disappeared.

Both the East Coast/West Coast jump rooms experienced losing power at times when their control room in Ohio was reporting normal operation of both jump rooms. The speculation was that the source of the jump room technology (Grey extraterrestrials) may have been covertly interfering with the functioning of the jump room teleportation technology, unbeknownst to control-room operators in Ohio. The U.S. government deployed Mendez, along with an evaluation team that included several prominent U.S. astronauts, to find out the causes of these discrepancies and the true destinations of the jump rooms. Basiago and Stillings have confirmed that Mr. Mendez was a federal investigator whose primary function was investigating questions like where the jump rooms were going.

SQEs

New Land in Time-Space?

Upon arriving at the jump room training class at College of the Siskiyous in 1980, Mendez debriefed Major Dames about his intelligence mission to evaluate the program. He then began a series of test jumps from the West Coast jump room. On one such test jump, he teleported with Mr. Obama and Mr. Basiago to a planetary environment that he initially thought might be the Mars they had been trained for. He recalls that when he had Mr. Obama shoot a flare at the sky, the flare bounced off a ceiling at about 62 feet up, indicating that they had teleported to an artificial environment or domed enclosure of some kind. The precise location of that enclosure was uncertain and may have been on a celestial body, such as Mars, or a freestanding holographic creation in space.

Mendez determined after multiple jumps that the destination the jump rooms were reaching was not Mars, which was out of position when most of the jumps took place. Its atmosphere lacked enough oxygen to sustain human life. Gravity on the surface was stronger than Mars's gravity would have been. The temperature swings on Mars (from -175°F to +225°F) would have made surface visits not survivable. Mars's lack of an ionosphere would have led to the irradiation of the jumpers. He concluded that the jump rooms were actually taking the participants to a "synthetic quantum environment," or SQE, a term invented by Basiago to describe a fold in the time-space continuum in which the Grey extraterrestrials have established an "artificial holographic planetary domain."[28]

During the ExoUniversity.org seminar, Basiago stated: "According to Bernard Mendez, the jump rooms were not visiting Mars but 'folds' in time-space created by the Grey extraterrestrials that the Apollo astronauts called 'slots.' These 'slots' or niches in time-space contain synthetic quantum environments in which the Greys have stored event scenarios that have taken place at different times on different planets, including Earth and Mars."[29]

The jump room program resulted, Mendez explained, from an ET-human liaison project involving the Greys and the U.S. government, in which the young Americans involved were selected by the Greys and then trained by the CIA. "This astonishing revelation means that President Barack Obama is a contactee selected by the Grey extraterrestrials to participate in the secret space program," Basiago commented.[30]

According to Mr. Mendez the U.S. government has identified 153 "synthetic quantum environments" constructed by the Greys in the near-Earth environment, ranging from 400 miles from Earth, on the surface of the Earth, and beneath the surface of the Earth. Because they contain new land that might be territorially acquired, the U.S. government was actively discovering and exploring these SQEs with a view to making them part of the United States.

Mr. Mendez was tasked to negotiate with the Greys as part of his evaluation mission. During the negotiations the Greys indicated that they had built the SQEs to be used for conditioning and educating humanity to be able to sustain a future role among organized intelligent civilizations in space and in the interdimensional multiverse.[31]

A Chrononaut's Contrary View

In the panel discussion sponsored by ExoUniversity.org held on June 2, 2012, that included all three whistle-blowers—Basiago, Mendez, and Stillings—Stillings shared his view that the destination the jump rooms were reaching was, in fact, Mars. He argued that the Martian moons Phobos and Deimos could be seen in their proper locations and trajectories in the sky and that jumper Courtney M. Hunt once stated that the Viking 2 apparatus was about twenty miles away from the jump room infrastructure that the jumpers called "the Corkscrew" because of its spiral structure. These clues indicated that the jumpers were on the real Mars and not a simulation.

NASA cannot be relied on to provide accurate data about Mars, Stillings stated, citing the space agency's claim that atmospheric pressure on Mars is only eight millibars. Such an atmospheric pressure could not have sustained the parachute deployment of the Mars Exploration Rovers Spirit and Opportunity to the surface.[32]

Basiago, who has spearheaded efforts to have America's chrononauts come forward and share their experiences with the public, called Mendez's account "a cosmic-level disclosure event." For his part Basiago thinks that Mendez and Stillings may both hold a key to the truth. He thinks that since the origins and operation of the jump rooms were unknown, they might have sometimes been taking the jumpers to Mars, but at other times were being diverted by the Greys to SQEs where human responses to the threshold of contact with the wider universe were studied.

In support of this explanation, Mr. Basiago cited the fact that Grey observation of U.S. activities in time-space was sometimes noted during time probes using U.S. time-travel technologies while he was serving on DARPA's Project Pegasus as a child in the early 1970s. He also cited the fact that power losses were being recorded during times when the jump rooms were ostensibly functioning normally. This could be evidence that during operation, the Greys were diverting the jump rooms from Mars to elsewhere.[33]

Basiago stated that since NASA has been lying about natural conditions on Mars, determining whether Mendez's "elsewhere" interpretation of the CIA jump room program or Stillings's "Mars" interpretation is the correct one will hinge on official declassification. He observed:

> What we have now is three fellow jump room participants sharing their experiences. . . . We were trained for Mars in summer 1980 and the domain that we were visiting from 1981 to 1983 was certainly understood to be Mars. What we are grappling with now is Bernard's claim that as the U.S. government investigator tasked to study the project he discovered that we were not visiting Mars but a . . . simulation of Mars architected by the Greys in a bubble Universe. . . . That's a highly provocative claim, but since Bernard Mendez and Brett Stillings and I were project participants together, in the interests of truth, I have facilitated all three of us coming forward publicly and sharing our understanding of what we were part of . . . Even if it was not Mars that we were visiting, all three of us agree that it was an off-planet location in time-space.[34]

Obama as a Time-Travel Pre-Identified CIA Asset

Whistle-blower Bernard Mendez's detailed testimony that he not only participated in the 1980–83 CIA jump room program with Barack Obama, but also engaged Mr. Obama in jump room experiments, such as a flare experiment to determine the location that the jump room technology was teleporting to, further bolsters emerging evidence that Barack H. Obama was involved at the early age of nineteen in a highly classified, top-priority U.S. national security project and went on to be a lifelong CIA asset or operative. It is logical, then, that the Obama administration would use such a high-level asset as Tommy Vietor, official spokesperson of the National Security Council, to deny, on January 3, 2012, that Mr. Obama was a participant in the 1980s CIA jump room program.

The CIA jump room program, and its origin with a species of Grey extraterrestrials with whom the U.S. government has had an ongoing secret human-extraterrestrial liaison program, is of the highest national security to the U.S. government. The CIA and U.S. national security apparatus surrounding President Barack Obama will do everything in its power to suppress evidence that Mr. Obama:

❊ Was part of a secret teleportation jump room program using Grey extraterrestrial technology to reach SQEs, destinations architected by Grey extraterrestrials

❊ Had been, since 1980, a full-time covert CIA operative, deployed to infiltrate, perform surveillance on, and report back on community activist, leftist, African nationalist, and Islamic circles

❊ Was pre-identified in 1971 via quantum-access time travel as a future U.S. president and had been essentially briefed and groomed for the job by the CIA since 1980*

Additional Eyewitness Testimony

The firsthand eyewitness testimony of Basiago and Stillings about the existence of a secret U.S. presence on Mars made possible by a revolutionary jump room technology is congruent with similar accounts given by three other Mars whistle-blowers:

❊ Former U.S. serviceman Michael Relfe, who spent twenty years as a member of the permanent security staff of a U.S. facility on Mars

❊ Former Lawrence Livermore National Laboratory scientist Arthur Neumann, who has testified publicly that he teleported to a U.S. facility on Mars for DoD project meetings

---

*Obama's identity as a CIA operative/asset and as being pre-identified as a future U.S. president is explored in my "How intelligence legend & Manchurian candidate Barack Hussein Obama was created," which can be viewed online at http://slidesha .re/18Im2fE.

✸ Laura Magdalene Eisenhower, great-granddaughter of U.S. President Dwight D. Eisenhower, who in 2007 refused a covert attempt to recruit her into what was described to her as a secret U.S. colony on Mars[35]

CORROBORATIVE MARS EXPERIENCES OF CAPT. RANDY CRAMER

In April 2014 U.S. Marine Capt. Randy Cramer came forward with his whistle-blower testimony claiming he had served seventeen years as a member of the U.S. defense force on Mars.[36] Following this, Mars experiencers Andrew D. Basiago and Randy Cramer as well as myself were part of an August 19, 2014, panel convened by Laura Magdalene Eisenhower to evaluate these Mars experiences.

Randy Cramer's Mars Report

Andrew D. Basiago states, "Let me enumerate ten elements of Randy Cramer's account that convince me that his account is real and not fictive":

1. Randy states that he is a captain in the U.S. Marine Corps. This is consistent with the secret military service that Randy is claiming. In the Army, Air Force, and Marines, captain is the first qualifying intelligence rank. It is the rank, for example, that an officer must hold to be given access to intelligence reports classified "Top Secret" or above. The corresponding rank in the U.S. Navy is Lieutenant Commander. This is the rank that Courtney M. Hunt of the CIA and Dr. Eldon A. Byrd of the Navy told me that I am listed in Navy records as having reached in order to document my service in Project Pegasus and the Mars Project. Randy's rank mirrors my own, despite the fact that he was attached to another branch of the

military and, in all likelihood, was unaware of my secret Naval rank because I had not widely revealed it.

2. Randy claims that he was born in 1970 and that he began training for Mars in 1987 at age seventeen. This is a young age to begin such dangerous service and tends to strain the credibility of anybody offering such an account. Seventeen was, however, the average age of the five Mars jumpers that have been identified as being among the ten American teenagers who were trained to go to Mars by Major Ed Dames at College of the Siskiyous in Summer 1980. They were: Barack H. Obama, born 8/4/61 (age nineteen), Andrew D. Basiago, born 9/18/61 (age eighteen), William C. McCool, born 9/23/61 (age eighteen), Regina E. Dugan, born 3/19/63 (age seventeen), and William B. Stillings, born 12/22/66 (age thirteen). Do the math. 19+18+18+17+13 = 85. 85/5 = 17. Clearly, the fact that Randy stated that his age at onset of training was 17 without knowing that this was the average age of our cadre of Mars visitors when we began training is probative of his claims.

3. Randy says he was part of the "Mars Defense Force." When we were training to go to Mars in 1980, Major Dames told our class that the three major reasons we were being sent to Mars were: (1) to help establish on Mars an extraterrestrial defense regime protecting Earth; (2) to create a basis under public international law to claim Mars as U.S. territory by sending temporary visitors and permanent settlers there; and (3) to acclimate Martian humanoids and animals to our presence. From 1981 to 1984 I took the jump room in El Segundo to Mars about forty times in furtherance of these goals. The fact that Randy states that he began training three years after my trips to Mars ended to participate in the establishment of a U.S. military presence on Mars is consistent with the developmental timeline. It would have been several years later that such a military buildup would

have begun. The fact that Randy was sent there not only as a Marine but as a settler harkens back to what Major Ed Dames told us about how Mars would be staffed during our training in summer 1980.

4. Randy claims that the ship that he traveled to Mars in used teleportation based on "wormhole propagation." I have given extensive public testimony that when the jump room morphed from a box into a cylinder halfway through our "jumps" to Mars, a wormhole, spiraling clockwise, formed at the far end of the cylinder. I have also described how my father, Raymond F. Basiago, who took one jump to Mars in May 1984, was the point man between Parsons and the CIA on teleportation, which remains a secret technical capability of the U.S. defense community. It would have been far easier for Randy to cite the space planes addressed frequently by sources within the aerospace community. But he didn't. He cited the far-harder-to-prove technology of teleportation and, in so doing, corroborated my teleportation account. How very interesting.

5. I have testified how I and my colleagues used teleportation to reach Mars via "jump rooms" that left an elevator shaft on Earth and after teleporting to Mars invisibly arrived in elevator shafts within U.S. facilities on Mars. Since Randy was a Mars settler, he could have imitated my testimony and stated that he, too, reached Mars via jump room. After all, we now have six whistle-blowers whose testimony supports the existence of jump rooms to Mars. They are: Michael C. Relfe, Arthur Neumann, Andrew D. Basiago, Laura Eisenhower, William B. Stillings, and Bernard Mendez. Laura even cited the acronym for the Mars jump rooms that we were given in our jump room training twenty-five years earlier. She was told during the recruitment effort directed at her in 2006–07 that she would go to Mars via "ARC," which stands for "aeronautical reposi-

tioning chamber." But Randy Cramer didn't ape the testimony of the six whistle-blowers that Mars is being reached via jump room. Instead, Randy testified that his teleportation was facilitated by advanced spacecraft that were more versatile in their movements than the jump rooms that we took to reach Mars. This makes perfect sense. As military men on Mars, Randy and his comrades would have needed such a spacecraft to provide a rapid response capability above the surface of Mars.

6. Randy states that the air on Mars is breathable. This was the previous testimony of myself and my fellow Mars jumpers William B. Stillings and Bernard Mendez. The fact that Mars's atmosphere is breathable has been established in photographs showing Mars's blue skies. I have testified as to Mars having blue skies, Randy as to the skies of Mars ranging from blue to purple in color. Oxygen is a clear gas that refracts blue. The Martian atmosphere is oxygen-laden as a result of oxygen seeping from its lithosphere, hence, its blue skies, now visible not only in Mars rover images from its surface but in satellite images showing the blue halo that envelopes the Red Planet like a thin azure canopy of gelatin. Case closed as to whether or not Mars is habitable. It is.

7. Randy describes how, on Mars, he was outfitted in a full body suit. The Mars jumpers were not; we were always dressed in our street clothes. The rust-red dust clinging to my hiking boots was once noted by my mother at our home in the San Fernando Valley and by a classmate sitting next to me at a lecture at UCLA.

8. Like me, Randy has described the U.S. facilities on Mars as "small." In light of the fact that our comparative testimony about this topic could have fallen into any one of nine combinations of large, medium, or small, it is conspicuous that our testimony is congruent and that we agreed without revealing our evidence to the other that the American facilities on Mars

are small. If descriptions of large, medium, and small were set as the parameters of our mutual descriptions of the U.S. buildings on Mars, then there was only a 13.71% chance that we would have randomly testified similarly as to this question.

9. I have since described how a member of the Martian species *Homo martis martis* once met Courtney Hunt and me at the entrance to the jump room facility called "The Corkscrew" and took us on a brief tour of the Martian underground, which was cavernous and decidedly "Tolkienesque."

10. In telling these details of his experiences on Mars, Randy Cramer strikes me as the very kind of individual that was selected for the Mars jump room program. He is highly intelligent, has a superb memory, is a composed and resilient individual, is spiritually strong and emotionally calm, and is highly articulate. The vocabulary that he demonstrated on Awake in the Dream included such graduate-level words as "psionically," "covertly," and "seditious"—not bad for a Marine when you're going to send him to Mars and ask him to report back.

Basiago gives a summation of his thoughts on the ten elements as follows:

I could not be more delighted by this development.

I salute Dr. Michael Salla for bringing Randy Cramer forward publicly as a Mars whistle blower. When Alfred Lambremont Webre brought me forward as a Mars whistle blower with a very similar account in 2009, Dr. Salla called me a "crackpot." Michael has now stated, almost apologetically, that in light of Randy's account, my account should now be reevaluated as credible. Let this begin a new era of collegiality in exopolitics.

I would also like to praise Randy's commanding officers for asking him to come forward. The fact that they have done so convinces me that my Truth Campaign has had a positive impact within the secret counsels of government.

Lastly, I would like to thank Randy Cramer himself. It took guts for Randy to live on Mars for 17 years and it has also taken estimable courage for him to come forward and tell of his experiences on Mars after he returned to the Earth.

Concluding that Randy Cramer's Mars experiences were congruent with and corroborative testimony of his own Mars experiences, Andrew D. Basiago wrote, "After careful analysis, I have concluded that the Mars testimony of Captain Randy Cramer is truthful. I am convinced that Randy is 'the real deal.'"[37]

As I stated during my recent appearance on Awake in the Dream Radio with Mark "Dr. Dream" Peebler and Laura Eisenhower, there are a number of highly specific facts contained in Randy's account that reveal to me that his account is authentic and that he is speaking from actual experience.

They are facts that are unlikely to have been made up and that would probably not be claimed by somebody purporting a false account. Some of them possess details that only a project insider would know. Others are facts that closely parallel the public testimony that I have given about my experiences visiting Mars for the CIA's Mars jump room program in the 1980's. They are also internally consistent when one considers Randy's account in isolation.

[For example,] Randy's description of blue bodysuits being worn by a phalanx of the Mars Defense Force corresponds exactly to a group of Martian humanoids dressed in blue bodysuits that I found in NASA-JPL image PIA10214 and included in my paper "The Discovery of Life on Mars" in 2008. . . .

Randy describes how the indigenous Martian civilization is "almost all underground." This is highly accurate. I have given dozens of radio interviews and public talks in which I have described the Martians as living "primarily underground." After I published "The Discovery of Life on Mars" in 2008, a senior career CIA officer telephoned me and informed me that the Martian underground civilization numbers about 1 million individuals consisting of several typologies living cooperatively.

The testimony of Captain Randy Cramer marks a signal event in the unfolding history of the Truth Movement and makes a major contribution to ending the cover-up of the fact that Mars is inhabited by both indigenous humanoids and animals and—now—also by visitors and settlers from Earth.[38]

Documentary (Photographic) Evidence of Martian Humanoids

The Mars rover program has provided the historic opportunity to photograph the surface of Mars. These photographs are transmitted to Earth and publicly released by NASA on the Internet. The public, including networks of Mars anomaly researchers, are then free to examine these NASA photographs for images of life-forms, including signs of current or past intelligent civilization, such as cities, structures, statues, sculptures, and so on.[39] The Mars Anomaly Research Society (MARS) is one such network of researchers who have published images of Martian humanoids and other Martian life-forms that have been located within NASA Mars rover photographs.[40]

Prima facie evidence demonstrates that Mars has a fragile, post-collapse ecology, inhabited by at least three species of intelligent non-Earthling humanoid life. In "The Discovery of Life on Mars," Basiago describes the *Homo martis extraterrestrialis* exophenotype as photographed by NASA rover Spirit:

Similar humanoid beings are present at other locations in this photograph taken on Mars. Inside the Rock Enclosure on Tsiolkovski Ridge, for example, a humanoid with two arms and two legs can be seen. He is kneeling with his back to Spirit. He has a bulbous head and an elongated body. He is wearing pants and a belt but is bare-chested. The scapulae in his upper back are evident. He is leaning over the wall away from the viewer. He may be reaching for, or lifting, a Martian child over the wall created by the back rock.

The humanoid being inside the Rock Enclosure on Tsiolkovski Ridge is easy to find in the photograph. Find the square in the upper left corner of the left solar array of Spirit, count five solar panels to the right on the edge of the array facing the valley, and proceed straight up the photograph. The C-shaped formation on the ridge is blue-gray in color. The humanoid is kneeling inside the enclosure of rocks. He is tiny and thin, like a praying mantis hatchling. He is interacting with other humanoids just beyond the rocks.[41]

The photographs Basiago is referring to are enlarged sections of NASA rover Spirit photographs PIA10214 and PIA11049. Basiago writes, "There is life on Mars. Evidence that the Red Planet harbors life and has for eons was discovered by the author by examining NASA photograph PIA10214, a westward view of the West Valley of the Columbia Basin in the Gusev Crater that was taken by the Mars Exploration Rover Spirit in November 2007 and beamed back to the Earth."[42]

It may be useful to the reader to examine photographs of Martian humanoid exophenotypes that have been discovered by the Mars Anomaly Research Society (MARS) founder Andrew D. Basiago and MARS researcher Patricio Barrancos of Argentina.[43] However, it is important to understand that NASA intentionally makes the images of Martian humanoid exophenotypes and other life-forms difficult to perceive and recognize in NASA rover photographs of the surface of Mars.[44] According to Basiago, "[MARS] has catalogued seven techniques that NASA is using to distort its data," as follows.

1. Scale. Compress vertical and horizontal dimension to make everything look ordinary.
2. Contrast. Set light-and-dark ratio to extreme value or invert light and dark aspects altogether.
3. Color. Add false hues and unnatural colors.
4. Skew. Move forms out of vertical alignment.
5. Consistency. Paint over to make a solid color look black, white, or opaque.
6. Integrity. Alter natural forms to look like forms that do not occur naturally.
7. Content. Embed content with altered contrast, non-human data, and strange pixels.[45]

Even with such distortion, at my public presentations of Martian humanoid exophenotypes, about 75 percent of the audience can recognize and see the humanoids in the photographs of Mars that are shown, as well as animals, statues, and structures (we poll the audience). At the November 2009 Barcelona Science and Spirit Conference, 50 percent of the audience that was polled could see. More recently, at public presentations in 2013, we have occasionally had up to 90 percent of the audience be able to perceive the exophenotype of *Homo martis extraterrestrialis* in a Mars Anomaly Research Society (MARS) photograph derived from a NASA Mars rover photograph of the surface of Mars.

Intelligent Life on Earth and Mars

This eyewitness and documentary evidence establishes, on the face of it, that it is more probable than not that intelligent life exists on Mars and that the U.S. government and other terrestrial governments now have outreach programs to the Martian civilizations as well as colonies on Mars. As we shall see, intelligent life on Mars relates to Earthlings not only through the time-space dimension, but hyperdimensionally as well.

4

Hyperdimensional Civilizations

As with the time-space dimension solar-system intelligent civilizations on Mars, both witness and documentary evidence indicate the existence of various species of intelligent hyperdimensional civilizations in our universe, each carrying out specific exopolitical agendas with respect to humanity. For example, from the platforms of the dimensions in which they are based, various hyperdimensional civilizations reportedly carry out "interdimensional teleportation" on large numbers and categories of humans.[1]

Eyewitness Evidence of Hyperdimensional Civilizations

As was mentioned in chapter 2, there is empirical evidence to show that a low estimate of 2 percent of the U.S. population and a high estimate of 14.7 percent of the total human population may have been physically, etherically, or mentally temporarily or permanently teleported to spaceships or dimensions outside our time-space Earth dimension by hyperdimensional civilizations.[2]

In 1991, [Budd] Hopkins, [Prof. David] Jacobs, and sociologist Dr. Ron Westrum commissioned a Roper Poll in order to determine how many Americans might have experienced the abduction phenomenon. Of nearly 6,000 Americans, 119 answered in a way that Hopkins *et al.* interpreted as supporting their ET interpretation of the abduction phenomenon. Based on this figure, Hopkins estimated that nearly four million Americans might have been abducted by extraterrestrials.[3]

The eyewitnesses to these teleportations report that they were carried out by various species of intelligent Grey hyperdimensional civilizations, as well as by other hyperdimensional civilizations. As mentioned earlier, expert estimates of the number of known Grey intelligent species range as high as 150 different species.[4] One study remarks on:

The "Hive Minds" of the [hyperdimensional] Greys and alleged abilities at "topological thinking" at speeds faster than the fastest computers known to mankind. The neurological superiority of Greys with discussion on the various cases, that specifically depict humans fearing Greys [per David Jacobs, Ph.D.] and Greys that have worked lovingly and positively with contactees, such as Suzanne Hansen [per Mary Rodwell]; Greys who have been seen with Anunnaki such as the case involving Futant* and Contactee Neil Freer.[5]

Replicable empirical research published by the International Community for Alien Research (ICAR)[6] confirms that a hyperdimensional civilization consisting of amphibian reptilian hybrids and Grey

---

*According to Neil Freer, "Futant is a term coined by Timothy Leary, Ph.D., former Harvard lecturer in psychology and the engine behind the peaceful Jeffersonian revolution of the '60's. Melding future and mutant, it defines those 1.5-2 percent of the population at any given time who are genetically 'programmed' to be the evolutionary future scouts."

cybernetic clones has reportedly abducted up to 14.7 percent of the present world population, or approximately 1 billion persons out of the current estimated human population of about 6,783,000,000 as of May 31, 2009. These human abductees, according to this research, have had their consciousness or their physical bodies taken by a quantum, hyperdimensional process and have then been returned to their original location (and in some cases original time) with memories of the abduction erased.[7]

Memories of hyperdimensional abduction retrieved through hypnotic regression and similar techniques reveal consistent patterns during the abductions, related to a covert agenda toward the human species by this hyperdimensional species. Because the abduction research data conflicts with the conventional paradigm of human reality, it is filtered out and denied, even by major researchers in the field of extraterrestrial studies. Cognitive dissonance or "the state of having inconsistent thoughts, beliefs or attitudes, especially as to behavioral decisions and attitude change" is one way of describing categories of responses to knowledge about and research into these hyperdimensional abductions.

ALIEN ABDUCTIONS AS PSYCHOLOGICAL EVENTS?

The canon of conventional academic science is that alien abductions are a psychodynamic artifact. One source states:

> The abduction phenomenon has garnered substantial attention from mainstream scientists and mental health professionals, who overwhelmingly doubt that the phenomenon occurs literally as reported and who have proposed a variety of alternate explanations, including "[d]eception, suggestibility (fantasy-proneness, hypnotizability, false-memory syndrome), personality, sleep phenomena, psychopathology, psychodynamics, [and] environmental factors."[8]

MILABs

Military abductions (MILABs) and psyops (psychological operations) abductions are a form of psychological warfare done by human military-intelligence agencies against targeted individuals for multiple purposes. One function of MILABS is to serve as human counter-intelligence in the case of extraterrestrial abductees. MILABS teams will abduct an individual who has already been abducted by extraterrestrials in an effort to determine why that individual is of interest to the extraterrestrials and thus to gain military and strategic intelligence about the motivations and agenda of the abducting extraterrestrial race.

The ICAR research confirms that alien abductions are largely done by hyperdimensional civilizations, and do not fit the category of military abductions. ICAR's research conclusions are backed by extensive, replicable research methodology and data, including objective correlation of answers to direct questions asked of Grey and amphibian-reptilian hyperdimensionals about the nature and purpose of alien abductions and their agenda toward the planet and humanity, by groups of twenty to fifty recurring abductees.[9]

Case Study

Hyperdimensional Warfare

A case study that illustrates how intelligent hyperdimensional civilizations strategize and operate is that of hyperdimensional warfare. This case study, which also may define the critical limits on any official extraterrestrial "disclosure" or open contact with extraterrestrial or hyperdimensional civilizations, may be deduced from eyewitness whistle-blower evidence of hyperdimensional armed conflict over control of our solar system.[10] If this apparent state of hyperdimensional "war" is factual (and not a false flag operation) and is part of a universal and galactic energetic transformation, then "official" extraterrestrial disclosure or peaceful open contact is likely to occur in the near future.

By a false flag operation, we mean a fictitious or mock war fabricated through holographic or other technology, or through mock battles between spacecraft, to trick human civilians into thinking there was a genuine war or invasion by extraterrestrial forces in space.

The eyewitness evidence includes whistle-blower testimony by two former U.S. armed forces personnel at a secret U.S. base on Mars. Their testimony describes actual scenarios in the dualistic military terms of defender and "enemy." As set out in the evidence below, the quantum-access nature of the "enemy's" arsenal of weapons systems suggests that the "enemy" may be a hyperdimensional or advanced third- or fourth-dimensional extraterrestrial civilization from another solar system or galaxy. The objective of this hyperdimensional "war" appears to be control of the third time-space dimension of our solar system.

WHAT IS A HYPERDIMENSIONAL WAR?

A conventional definition of war is:

> a behavior pattern exhibited by many primate species including humans, and also found in many ant species. The primary feature of this behavior pattern is a certain state of organized violent conflict that is engaged in between two or more separate social entities. Such a conflict is always an attempt at altering either the psychological hierarchy or the material hierarchy of domination or equality between two or more groups. In all cases, at least one participant (group) in the conflict perceives the need to either psychologically or materially dominate the other participant. Amongst humans, the perceived need for domination often arises from the belief that an essential ideology or resource is somehow either so incompatible or so scarce as to threaten the fundamental existence of the one group experiencing the need to dominate the other group. Leaders will sometimes enter into a war under the pretext that their actions are primarily defensive; however, when viewed objectively, their actions

may more closely resemble a form of unprovoked, unwarranted, or disproportionate aggression.

In all wars, the group(s) experiencing the need to dominate other group(s) are unable and unwilling to accept or permit the possibility of a relationship of fundamental equality to exist between the groups who have opted for group violence (war). The aspect of domination that is a precipitating factor in all wars, i.e., one group wishing to dominate another, is also often a precipitating factor in individual one-on-one violence outside of the context of war, i.e., one individual wishing to dominate another.[11]

HYPERDIMENSIONAL "WAR" AND QUANTUM-ACCESS TECHNOLOGY

The possible hyperdimensional "war" has come to public light because of the operational use of quantum-access time-travel jump gates and quantum-access technology at the U.S. secret base on Mars. Quantum-access technology is used to time travel U.S. armed service personnel who serve as permanent staff at that base and to age-regress them back to their original time-space starting date after completion of their 20-year rotation cycle on Mars. The base thus may be hyperdimensionally "cloaked" from civilian society on Earth and hidden in covert niches in the time-space hologram. As described by former U.S. chrononaut Andrew D. Basiago, one early application of time-travel technologies was the hiding of national security secrets from the Soviet Union during the Cold War, for example, by time traveling these secret documents to quantum-access niches in the time-space hologram.[12]

EYEWITNESS EVIDENCE OF WARFARE WITH HYPERDIMENSIONALS

Two independent whistle-blowers, both former U.S. armed-service personnel stationed at the secret facility on Mars, have now come forth with detailed accounts of their experiences in warfare with hyperdimensional intelligent civilizations.

Michael Relfe is one such whistle-blower. A former member of the U.S. armed forces, he was recruited as a permanent member of the secret U.S. facility on Mars. In 1976 (Earth time) he was teleported to the U.S. Mars facility and spent twenty years as a permanent member of its security staff. In 1996 (Mars time) Mr. Relfe was time traveled via teleportation and age-regressed twenty years, landing back at a U.S. military base in 1976 (Earth time). Mr. Relfe then served six years in the U.S. military on Earth before being honorably discharged in 1982.[13] He is a credible whistle-blower whose life and career have suffered greatly because of his decision to inform the public of the secret war against hyperdimensional intelligent civilizations.[14]

In a two-volume book authored by his wife, Stephanie Relfe, *The Mars Records*, Relfe describes the two types of individuals at the secret Mars colony:

> There are two kinds of people that I remember.
>
> 1. People visiting Mars temporarily (politicians, etc.)—They travel to and from Mars by jump gate. They visit for a few weeks and return. They are not time traveled back. They are VIP's. They are OFF LIMITS!!
>
> 2. Permanent staff—They spend 20 years' duty cycle. At the end of their duty cycle, they are age reversed and time shot back to their space-time origin point. They are sent back with memories blocked . . . to complete their destiny on Earth.[15]

Relfe also discusses the presence and functions of reptilian and Grey extraterrestrials at the secret Mars colony. Below, Relfe is questioned under the direction of his wife Stephanie Relfe using memory retrieval technology.

EL: What about the Reptilians?

MICHAEL RELFE: Yes. They are racially related (Draconians, Reptilians, Greys).

EL: Do any Greys and Reptilians live on the Mars Base?

MICHAEL RELFE: Yes, some are stationed there. I remember the Greys as doctors or technicians. I believe the Reptilians stay camouflaged (cloaked) most of the time. They prefer to appear human because they are naturally fierce-looking.[16]

The following excerpt from *The Mars Records* illustrates the nature of the armed conflict occurring around the U.S. secret base on Mars and the types of weapons used. Relfe describes the secret Mars base as being under attack by a hyperdimensional "enemy" that apparently can time travel. This discussion begins with a reference to the "time-travel jump gates," the portals used for time-travel age regression of permanent staff of the base to return them to their original time-space point of departure from a U.S. military intelligence base on Earth.

I was not a jump gate technician so I only remember things from a layman's point of view. This technology was a result of the Philadelphia Experiment projects as described by Dr. Al Bielek. It's one of those things that are "Ohh, Wow, Fantastic" the first time you see them, and then they become taken for granted. I also remember that they were guarded REALLY tight and that every moment of use was accounted for and logged.

I remember that there were several jump gate stations on [the] Mars Base. These stations were "hooked to" other places and they were defended against someone or something. I remember that RV technicians hooked to the machines assisted in that defense. Remember that Tactical Remote Viewing is not only used to terminate targets. Termination is a tiny percentage of all operations.

It is used mostly to defend against the enemy. It is used to defend Very Important People (VIP's), Very Important Equipment (VIE's), areas and regions of the planet. In addition, some weapons systems require an RV operator to monitor and direct them. There are physical weapons systems (example like a particle beam projector) and there are other types of weapons that are energetic (Psi) types of systems. Also RV operators are commonly assigned to monitor attack craft on patrol.

The enemy also has Psi types of weapons systems that they direct against our physical type static shielding. Static shielding doesn't change. It stays the same no matter what you throw at it. To augment the static shielding, you have RV operators that monitor and probe the shields looking for penetration. If penetration is discovered, an RV operator can trace the source through space-time and deal with the intruder. RV operators on duty are not alone. They have backup operators in addition to supervisor operators, they can instantaneously get backup if the need arises. I have seen many ads for books and courses for remote viewing. Most of what passes for RV at this time is about the first 3 or 4 days of training at [the] Mars Base. And most people involved in this sort of thing would, unfortunately, never survive the training. The drugs would kill them quick. Referring to the drugs, Michael Relfe states, "The drugs were used to stimulate and 'open up' the metaphysical abilities of the candidate. I do not know the names of them. I remember them being gentle and interesting at first but quickly the effects became hard and unpleasant on the body."[17]

In one passage Relfe recounts an "attack run" he undertook on "an enemy ship" where he lost consciousness when the enemy missile penetrated his ship. He "was flying solo in a space ship when he encountered an enemy craft. Instead of fleeing he went for the enemy craft. It fired a missile at him, which did not explode his ship, but it did penetrate his ship and go through him. This caused incredible damage. He would have died had he not had a special suit which was programmed to administer first aid."[18] Relfe describes the advanced technologies possessed by the hyperdimensional "enemy."

> The [hyperdimensional] enemy has at their disposal multiple technologies that will allow them to abduct an individual from their bedroom, examine/operate/indoctrinate them, and return them to their bedroom without being detected. They can "keep" an individual for weeks and return them to a time a few minutes after they were taken. The names of those technologies that I am aware of are:

> Jump gate technology
> Fractal Jump gate technology
> Transporter technology
> Fractal Transporter technology
> Teleportation technology
> Fractal Teleportation technology
> Time Travel technology
> Fractal Time Travel technology
> Dimensional Travel technology
> Fractal Dimensional Travel technology
> Wormhole Travel technology
> Fractal Wormhole Travel technology
> Resonance technology
> Fractal Resonance technology
> Magical technology
> Fractal magical technology
> Walk-through-the-walls technology
> Fractal walk-through-the-walls technology
> A "sender" sending something into your room
> A "sender" pulling you out of your room[19]

WHAT IS THIS HYPERDIMENSIONAL "WAR" ABOUT?

The dimensional nature of the "enemy's" arsenal of weapons systems suggests that the "enemy" may be a third- or fourth-dimensional extraterrestrial civilization. There are other possible interpretations of the eyewitness evidence about whether there actually is a hyperdimensional "war" for control of the third dimension of our solar system. For example, who is the "aggressor" or "invading party" in this hyperdimensional "war"?

First, Earthling humans may be the aggressor. The secret U.S. facility on Mars that itself uses jump room technologies may be perceived by the hyperdimensionals as an aggression by Earthling humans, inviting counter-attack by the hyperdimensional "enemy" at the time-travel jump gates of the Mars base.

Second, the hyperdimensional "enemy" may be the aggressor. The U.S. facility on Mars may be a strategic forward position by the Earthling human forces in alliance with the Martian humans (living under the surface of Mars) to ward off a threatened occupation of the Mars third time-space dimension by the hyperdimensional "enemy."

Third, the "armed attacks" by the hyperdimensional "enemy" on the U.S. base on Mars may be part of an elaborate false flag deception scenario to create a faux "war" against the "evil extraterrestrials."

Of these alternatives it seems most reasonable to conclude that the "hyperdimensional war" around U.S. facilities on Mars as reported by Michael Relfe is accurate and is an example of hostile, competing interaction by intelligent civilizations based in other dimensions.

Archons

Hyperdimensional Controllers

The ancient Gnostic texts from Egypt, the Nag Hammadi library, describe two types of demonic alien beings that invaded Earth long ago, which they call the *archons*. The first type of archon looks like a reptile, the other like a human embryo, with the same shape and appearance as the "sky fish" photos.[20] To sort out and clarify what the Sophianic narrative may have to say about the test of the archons is a great challenge to our understanding of the Gnostic message and how it can benefit humanity today.[21]

According to researcher and cosmologist Laura Magdalene Eisenhower, archons are hidden negative controllers of humankind; they are inorganic interdimensional entities that must now be exposed and exorcised from the individual human mind, from our human species, from the planet, and from our universe as a whole, as part of our collective evolution to a new state of consciousness and being.

Researcher Robert Stanley maintains that humanity must now take a scientific approach to identifying archons and exterminating them.

It is time to expose the covert controllers of mankind. I assure you this is not speculation, a hoax, or the figment of peoples' imagination. These parasitic creatures are real and they need to be dealt with immediately so mankind can evolve to the next level of existence.

Although these parasites are not human, they feed off the negative energy/emotions of humans. It is unclear when these cosmic, amoeba-like creatures first came to Earth, but we know they were discovered by shamans in altered states of consciousness long ago and have recently been photographed. The reason everyone is not seeing them on a daily basis is because the creature's energy signature is beyond our normal, narrow range of vision within the electromagnetic spectrum, what scientists call "visible light."[22]

In discussing archons John Lash[23] writes:

Although archons do exist physically, the real danger they pose to humanity is not invasion of the planet but invasion of the mind.

The archons are intrapsychic mind-parasites who access human consciousness through telepathy and simulation. They infect our imagination and use the power of make-believe for deception and confusion. Their pleasure is in deceit for its own sake, without a particular aim or purpose. They are robotic in nature, incapable of independent thought or choice, and have no particular agenda, except to live vicariously through human beings. They are bizarrely able to pretend an effect on humans, which they do not really have.

For instance, they cannot access human genetics, but they can pretend to do so, in such a way that humans fall for the pretended act, as if staged events were taken for real. In this respect, archons are the ultimate hoaxers. This is the essence of "archontic intrusion," as I call it. The trick is, if humanity falls under the illusion of superhuman power, it becomes as good as real, a self-fulfilling delusion.[24]

Mary Rodwell

*Hyperdimensional Intelligent Civilizations
Interact with Earthling Humans*

The dimensional ecology of the Omniverse hypothesis is supported by replicable empirical evidence gathered by the Australian Close Encounter Research Network (ACERN) and researcher Mary Rodwell, who has developed an evidence-based model of how hyperdimensional intelligent civilizations interact with Earthling humanity in the Omniverse. Her model demonstrates how hyperdimensional civilizations communicate ethically and cooperatively with humans at an individual and collective level.[25]

In her article "Extraterrestrials, human consciousness, and dimensions of soul: The intimate connection," Mary Rodwell presents her research into more than 1,800 cases of abduction by teleportation of contactees by hyperdimensional civilizations in the dimensional ecology. She concludes that the hyperdimensional extraterrestrials, the human subjects, and their abductions potentially demonstrate "a new reality: a multiverse."[26] She writes:

> The human connection to extraterrestrial intelligences: does it transcend space and time? I believe we are trying to explore a multiverse through the eye of a needle.
>
> The majority of the public and most researchers believe that Contact or encounters with extraterrestrials are defined through two, very limited windows of Contact experience. Firstly, the Abduction scenario, when the individual may feel paralyzed, and taken in a trance, a deep sleep-like state on-board craft. Secondly, the contactee, who has full conscious recall of their Encounter. My research, however, suggests a third, far broader category, which is far more common and less physically intrusive. It encompasses some of the above, but embodies more subtle patterns of experience. If we broaden the mandate to be inclusive of this third category, we stand to gain a far more fascinating picture of Contact and how or why these intelligences are interacting with us.
>
> In this article I explore all forms of Encounters and ask why research with Experiencers of Contact with these intelligences offers such diverse and conflicting data. What approach is the best for the Experiencer seeking answers and verification of [such] encounters? I explore the benefits of the holistic approach to gathering data.
>
> [The article sets out] misconceptions regarding data revealed by hypnosis, and [questions if this] extraordinary data is confabulation, fantasy, or potentially demonstrating a new reality: a multiverse.
>
> The testimony from Experiencers challenges not only consensus reality, but our dearly held religious and spiritual beliefs, and demonstrates an intimate and complex association with extraterrestrials, a connection that spans both space and time, including the realms and dimensions of soul.[27]

Midway through her article Mary Rodwell goes on to explore the multiverse as well as the data supporting interactions between various dimensions in this multiverse. She writes:

This phenomenon asks us to be courageous and challenge consensus reality, to be "wise" enough to accept what we don't know, cognizant that our present understanding of reality is being framed by a science in transition. Concepts such as quantum reality, the holographic universe, unity consciousness, [are] theories which suggest that everything is interconnected through consciousness. This is a very similar concept to what those with Encounters convey, but experientially. Their experiences demonstrate this alternate, multifaceted paradigm that could prove to be the "true nature of reality." I feel that we as a species are being challenged by such Contact to relinquish our old reality programs. Encounters [are] a modern-day "shamanic experience," designed to break down outmoded programs of reality and allow for us to become fully conscious and operate in a multidimensional reality. In Awakening I discuss this as the "realization event," when the person finally accepts that the old paradigm no longer works as they embrace the new one. It can be terrifying to have all your beliefs of reality challenged and to accept an expanded reality. But like the apprentice shaman, they face their ultimate fears, to explore and cope with a multidimensional perspective of reality. The shamanistic experience was found to be parallel in outcome to Encounters, bringing transformative, heightened psychic and healing abilities to the individual. . . .

But, if true, [these data] indicate interactions/encounters not only in physical form, but when we are separated from our physical body, such as in an astral or out of body (OBE) soul state. But added to this seem to be interactions through many lifetimes, and the in-between-lives and before our present human incarnation, encounters that continue after our present physical life ceases. So to all intents and purpose the interactions are in both physical and nonphysical [forms] through many timelines, as we are educated and assisted in physical realms and the realms and dimensions of the soul.

This information is profound and answers . . . one of the most pressing of questions asked by the Experiencer, especially [those who] have felt victimized by their encounters. To their astonishment, when the question regarding consent is asked, they may be shocked to discover this permission was given, before their present human incarnation. They describe themselves as an intelligence, which appears to them to look like a bright orb of light, their soul essence. They vocalize: "I agreed, but it was before I came here." Some add, "I have a job to do. To assist in the evolution in human consciousness." Hundreds have verbalized such statements.

I recall sharing this information, with one well-known author and fellow researcher. He totally dismissed the data, and argued this information is a form of delusion, a program instilled by these Intelligences hoping to trick the individual into accepting their encounters. I find that explanation very difficult to believe, given how my clients resonate at some deep level, the fact that it makes so much sense to them. I question who is paranoid. Do we trust or negate what resonates to the Experiencer, or what we as observers prefer to believe? I do not have the answers, I do not pretend I have, but I do have concerns when anyone articulates such a mandate, without any data to prove such a negative and fear-based perspective, [it] is unhelpful to this research.[28]

Ms. Rodwell quotes hyperdimensional experiencer Tracey Taylor, of western Australia:

As far as my own contact experiences I became very compassionate towards the ET beings as there is much humans can learn and benefit by interacting with them. I was fearful of my experiences only when I realized that I seemed to be the only one having them. I didn't know what was going on or why. But there has always been an "equal" exchange. I helped them achieve their genetic goals and in return my healing and psychic ability, plus my life on earth and so on has been enhanced. They also protect me when I ask them to. Basically at the start I misunderstood the Greys. The knowing "ET" part of me initially made a decision to assist them and understanding of this can be overshadowed by fear which stems from our limited human perceptions and reactions to these experiences.[29]

About the "soul domain," Mary Rodwell writes:

If we are open to what we don't know, it leads us into some fascinating hypotheses. [Consider] the OBE encounter, where many individuals perceive themselves as "orbs" or spheres of light. They describe leaving their body as an "orb of light," in metaphysical terms called a *merkabah*. This happened to Experiencer Peter of Sydney, Australia, who stated, "A beautiful space, I love it, it is my home, we are everything, I'm every shape, I was a born soul, just beautiful, with energy, the energy of the soul, we create, you create, I'm liquid, I'm fluid, I can go here or there, like a bubble, this is the form of soul.". . .

The "orb" of light is suggestive of the "soul state" when traveling out of body. In some encounters, people relate that their partner cannot be woken up, no matter how they called out to them, shouted, or shook them. In regression hypnosis, I had one explanation. The "awake" person perceived an orb of light entering the bedroom a bright, greenish "orb." It moved about the room, and it was so strange and scared the person. Very soon after it appeared, the sleeping person awoke. In regression this orb of light proved to be the astral body or OBE "soul state" of the sleeping person, returning to [the] physical body.

In another instance, it was demonstrated that we may inhabit other "life-forms" while in this OBE state. An Experiencer, one of what I call a Star seed, felt she had "dual consciousness," articulated as part human, part ET. In regression, she described being on-board a craft, saw herself in the form of a Grey being, working as a scientist. I asked where her human form was located. She answered, it was on the craft but inanimate, while her consciousness inhabited her "ET self." I asked how she returned to her human body. She described a ball of light leaving her extraterrestrial form to re-animate her human body. She said it was like she changed one biological overcoat for another. Many other individuals have experienced this swapping of physical form to an extraterrestrial form when in an OBE state.

This research offers an insight that demonstrates an extraordinary perspective of our intimate connections to these intelligences, and invites more questions as to the true nature of reality. It examines concepts such as interspecies genetic links, individuals told they may have DNA from one or more extraterrestrial life forms, links that transcend space-time, and interactions with souls before and after human incarnation. Some extraordinary data came from a 15-year-old boy I worked with in the UK in 2008, when on a lecture tour. The data was intriguing, especially given the age of the young man. There were many fascinating aspects to the case, and I did hypnotic regressions with two members of a family of four. They had a joint encounter and experienced "missing time." I regressed the grandmother, which was extremely successful in dealing with her long-held fears, fears she finally released through this process. Her 15-year-old grandson I will name John (not his real name). John was very uncertain he wanted to have hypnosis, fearful of what he would discover. However, the positive, healing outcome for his grandmother encouraged him to proceed. In the regression he spoke about the profound nature of what he termed "new souls." We have to be reminded this was a 15-year-old boy speaking about concepts he had no awareness of consciously. In one encounter with the Greys, he also recognized his departed relatives, describing them appearing as "orbs" of light. He said they were there to support him, to stop him being frightened.[30]

It is certainly appropriate for us to ask why the Earth's time-space dimension is so important to the hyperdimensional intelligences. Mary Rodwell concludes:

A complex extraterrestrial program, genetic links, multilevel educational programs, coupled with soul interaction, is not congruent with a subversive agenda. Even to my limited human logic, that is nonsensical. Extraterrestrials have technology to take us out in a nano-second if that was their desire. No, these encounters suggest a far different hypothesis, something far more profound, and the reward is transformation of consciousness.[31]

5

Intelligent Civilizations' Governance Authorities

The Omniverse is, by the evidence, highly populated. Please recall that Stanford physicists Linde and Vanchurin have calculated that there are 10 raised to the power of $10^{10,000,000}$ physical universes in the multiverse, and that a highly conservative estimate of the number of intelligent civilizations in our universe alone is 100 billion (an estimate based on the assumption that there are only twelve intelligent civilizations in our galaxy, whereas we know now there are at least 50 million Earth-like planets).[1]

The Western official governmental and academic culture at present assumes that Earthly institutions and actors are the only stakeholders in outer space.[2] However, there is presumptive evidence that the multiplicities of intelligent civilizations in the universes of the multiverse are organized via governance authorities. While the database of evidence for these governance authorities is in its early stages of development, it will improve as disclosure of official extraterrestrial monitoring and liaison programs by governmental military and intelligence agencies, such as those of Russia, goes forward, and as information that Earthling governments have developed is released into the public domain. Moreover, once open public contact and interaction with intelligent civilizations commences officially, these civilizations will share more information about their governance authorities and governance in our larger universe, as well as in parallel universes of the multiverse.[3]

Nature of the Evidence for Existence of Intelligent Civilizations' Governance Authorities

As of this writing there are at least three sources in the public domain of probable exopolitical evidence for legally constituted intelligent civilizations' governance authorities that exist in the multiverse, which have jurisdiction over defined dimensional territories, such as the Milky Way galaxy. This probable evidence includes types of eyewitness and documentary evidence that are within the protocols of evidence acceptable to the science of exopolitics. Three specific sources are:

❧ Scientific remote viewing

❧ Contactee, telepathic, and documentary evidence

❧ Information collected by Earthly governments from direct interaction with intelligent civilizations

SCIENTIFIC REMOTE VIEWING

Communication with representatives of governance authorities of our Milky Way galaxy, using standard laboratory remote-viewing protocols, has been reported to the public.[4] According to the Farsight Institute, remote viewing is:

> a controlled and trainable mental process involving psi (or psychic ability). Remote-viewing procedures were originally developed in laboratories funded by the United States military and intelligence services and used for espionage purposes. The scientific understanding of the remote-viewing phenomenon has greatly advanced in recent years, and as a result the process of remote viewing can now

be reliably demonstrated in both laboratory and operational settings. There are a number of styles of remote-viewing procedures that are popularly practiced, such as Scientific Remote Viewing (SRV), Controlled Remote Viewing (CRV), Hawaii Remote Viewers' Guild procedures (HRVG), as well as a few others. Remote viewers use one or more of these styles to gather descriptive data of a "target," which is usually some place or event at some point in time.

Remote viewing is always done under blind conditions, which means that the remote viewer must know nothing about the target when conducting the remote-viewing session. All of the various styles of remote viewing require both training and regular practice in order for a remote viewer to become proficient. Remote viewing is normally considered a controlled shifting of awareness that is performed in the normal waking state of consciousness, and it does not typically involve an out-of-body experience, hypnosis, an altered state of consciousness, or channeling.[5]

Remote viewing is a consciousness technology involving a standard laboratory protocol that can access intelligences in nonlocal subspaces in the multiverse and produce replicable communication. Remote viewing thus provides a means to transmit and receive scientifically replicable information between a trained individual based on Earth in our time-space hologram and a representative of an intelligent civilizations' governance authority in some adjacent dimension of our universe or parallel universes in the multiverse. Remote viewing is, therefore, an ideal laboratory protocol and tool for support and verification of the dimensional ecology of the Omniverse hypothesis.

Replicable data from an experiment by scientific remote viewers associated with the Farsight Institute suggest that a spiritually and technologically advanced galactic federation of worlds exists. It can be described as a sort of loosely organized dimensional government of our Milky Way galaxy.

For example, the following is a transcript of a scientific remote viewing session with a galactic federation leader in the Milky Way:

REMOTE VIEWER: As soon as I went into his mind [that of a galactic federation leader], I re-emerged in space. That is where I am now. I am outside of the Milky Way, looking onto it. Dotted lines have been drawn over the image, dividing up the galaxy-like quadrants.

I am being told that there is a need for help. They need us. I am getting the sense that they need us in a galactic sense, but I seem to be resisting this. They are so much more powerful than humans; it just does not make sense why they would need us.

The leader is sensing my resistance and redirecting me to a planet. OK, I can see it is Earth. I am being told that there will be a movement off the planet in the future for humans. I am just translating the gestalts now into words. But the sense clearly is that Earth humans are violent and troublesome currently. They need shaping before a later merger. Definitely humans need to undergo some sort of change before extending far off the planet.

MONITOR: Ask if there are any practical suggestions as to how we can help.

REMOTE VIEWER: I am being told in no uncertain terms that I am to complete this book project. Others will play their parts. There are many involved, many species representatives, groups.

MONITOR: Ask who else we should meet using remote viewing, or another technique.

REMOTE VIEWER: Only the Martians. Hmmm. I am being told that our near-term contact with extraterrestrials will be limited to the Martians for now, at least in the near future.

MONITOR: Ask if there is new information that we need to know but do not know.

REMOTE VIEWER: This fellow is very patient. He knows this is hard for me. He is telling me that many problems are coming. There definitely will be a planetary disaster, or perhaps I should say *disasters*. There will be political chaos, turbulence, an unraveling of the current political order. As we are currently, we are unable to cope with new realities. He is telling me very directly that consciousness must become a focal concern of humans in order for us to proceed further.

He is right now tapping into *your* (my monitor's) mind. It is like he is locating you, and perhaps measuring or doing something. He is telling me that you are very important in all of this. We must come back here—their world—at later dates. We will be the initial representatives of humans as determined by consciousness. He is telling me that consciousness determined our arrival point. There is more. We are not saviors, just initial representatives. He wants me to get this straight.

I am getting the sense that he wants us to understand that we have a responsibility to represent fairly. This is not to go to our heads. This is just our job now, and we all have jobs. He is also telling me that I am doing a fairly good job at writing this down.

He likes your sense of humor. He says that there will be *lots* of activity in the future, of the day-to-day sort. But, for now, we are to focus on the book. The book is important, and they will use it.[6]

CONTACTEE, TELEPATHIC, AND DOCUMENTARY EVIDENCE

The details contained in evidence of eyewitness contactee and telepathic interaction with representatives and intermediaries of reported intelligent civilizations' governance authorities vary widely and are not necessarily consistent with each other or with known history. More than anything, this tends to be evidence, on the face of it, that such authorities do exist, rather than evidence of the content of their mission, role, and history. Wherever possible, the contents of the prevailing evidence on governance authorities are checked for congruency with comparable information.

INFORMATION COLLECTED BY EARTHLY GOVERNMENTS FROM DIRECT INTERACTION WITH INTELLIGENT CIVILIZATIONS

Accurate and reliable information collected by Earthly governments from direct interaction with intelligent civilizations through teleportation, time travel, and other space travel missions into our universe and the multiverse will occur upon truthful and complete governmental disclosure.

I have interviewed several whistle-blowers from secret U.S. extraterrestrial liaison programs that have involved teleportation and time travel, including former U.S. chrononauts Andrew D. Basiago, William B. Stillings, and Bernard Mendez, former U.S. serviceman Michael Relfe, and others. Research is ongoing into issues of intelligent civilizations' governance and will be reported as it is developed.

Role of Governance Authorities in the Origin of the Human Race

Reported telepathic communications with representatives of regional governance councils in our galaxy have provided evidence of the possible origins of the Earthling human race that is congruent, at first sight, with scientifically derived information about human DNA.

For example, one telepath who is in contact with a reported regional galactic governance council states that a consortium of advanced intelligent upper-dimensional extraterrestrial and interdimensional civilizations originally developed *Homo sapiens* as an intelligent being with twelve-strand DNA, as a species that was to have been a guardian of the third dimension of time-space on Earth.

In discussing the development of *Homo sapiens* by a consortium or council of advanced, ethical, intelligent hyperdimensional civilizations, including Pleiadians and Sirians, telepath Patricia Cori states:

Intensive studies were made of the existing environments of Earth—the distinctive plant and animal kingdoms—and detailed investigations were conducted into how biodiversity resulted as a reflection of various geographical and climatic variables. It was discerned that such diversity would provide ideal conditions for the seeding of extraterrestrial species, as their original environments could, in many ways, be replicated in the ecosystems of Earth.

Unexplained in your archeological and missing-link evolutionist theories is the isolated emergence upon your planet of four distinct seed races. These are archetypes of master races whose fundamental genetic material formed the primary "substance" of your race, while the vibrational patterns and sequencing of extradimensional beings (those of higher realms) were woven into the intricate light codes of your incredible twelve-stranded DNA. The genetic material of these four primary races [black, red, yellow, and white] was united into the blueprint of the species of Homo sapiens.[7]

Multiple Reports of a Regional Galactic Governance Council

Patricia Cori's reports of telepathic communications with a regional galactic governance council are congruent with recent empirical data and exopolitical analysis indicating the existence of such a regional galactic governance council, concerned with Earth's ecology and humanity. There is also persuasive documentary evidence of intelligent civilizations' governance authorities, consisting of photographs and video recordings of decloaking of these authorities' spacecraft over major Earth cities on predicted dates.

Former Canadian North American Aerospace Defense Command (NORAD) officer Stanley A. Fulham has reported interacting with representatives of a regional galactic governance council whose membership includes upper-dimensional, ethical Pleiadian and Sirian civilizations, who are concerned with the development of life-bearing planets in this area of our galaxy, including Earth.[8] It is important to know that Fulham was a lifelong NORAD officer who directed the NORAD radar that tracked hyperdimensional ET UFOs across and around the North American continent.

When I interviewed Fulham he stated that a regional galactic governance authority decided in January 2010 to put aside the law of nonintervention. At a solemn meeting the regional authority decided to intervene with their technology to clean Earth's atmosphere before an environmental collapse occurs on Earth, as has happened on many other inhabited planets with civilizations similar to our own. The authority did so, according to Fulham's information, after concluding that our human technology could not prevent an environmental collapse and species extinction on Earth.

According to Fulham this is a rare decision by the authority, resulting partly because authority members wish to preserve the "unique positive qualities of our human population." The authority, according to Fulham's information, "consists of the intelligent civilizations of the Pleiades, Orion, Sirius, Bootes, Alpha Centauri, Comsuli, Zeta Reticuli, and Pouseti."[9]

The regional galactic governance council, according to Fulham, has had a caretaker role for our planet for about the last million years and has effectively maintained Earth under protective quarantine following an attempted invasion and takeover of Earth.

STANLEY A. FULHAM'S DECEMBER 3, 2010, STATEMENT

Just prior to his untimely death, Fulham published the following statement on his website:

> I had advised my readers and UFOlogist supporters I would notify them by e-mail on the next major UFO display in their "intervention process" with mankind which started in their major display

over New York on 13th October 2010. The prediction, documented in my book *Challenges of Change** . . . stated that "there would be a major display over our principal cities" on that date.

There was a great deal of speculation about this change in the Transcendors prediction. ["Transcendors" are Fulham's interdimensional telepathic connection to the regional galactic governance authority.] In late September 2010 I asked the Transcendors about this significant change. They replied: "The Pleiadians are much more aware of the consciousness of mankind than you are—after all, they have been observing you for millions of years. They came to the conclusion that mankind (generally speaking) is not prepared for a massive display of their UFOs over our principal cities. They would regard it as a hostile intervention—a prelude to invasion. Accordingly, the Alien Council decided about three weeks before their 'display' they would focus on one of our principal cities, New York. It is recognized worldwide as our most cosmopolitan city with a population not likely to be overwhelmed by the sight of UFOs; however, the 'display' date of 13 October 2010 would remain." There were other major UFO sightings on that date all over the world. But the Transcendors caution that they were not part of the Council's principal focus over New York. The New York display is simply the beginning of the Aliens' "intervention process."

In my last "Reading" with the Transcendors on Tuesday, October 13, 2010, the Transcendors advised me of the following predictions on the Pleiadian UFO displays:

A. There will be a major UFO display over Moscow between the end of the first part of January 2011 and the second week of January 2011.

---

*Fulham's *Challenges of Change* is the product of ten years of research and interaction with a council-connected source and contains a blueprint of the council plan for 2010 to 2015.

B. This "display" will be followed by a major display over London approximately seven days later.

C. Interventions will then accelerate, not so much over our cities, but dispersed over our continents, with sightings increasing in duration. The intent of these interventions is to increase mankind's acceptance of the alien phenomena, so that, hopefully, we will be prepared to accept a face-to-face encounter and communicate, perhaps as early as next year (2011).

D. Objective—a dramatic introduction of the alien reality—an appearance at the United Nations.

The Alien "intervention process" will provide a philosophical challenge to mankind as nothing else has ever done. It will challenge our most fundamental beliefs. Did Christ teach to this extraterrestrial world? Are they aware of Allah and his hatred for the heathens and infidels (which, of course, must include the aliens—they also have souls!)? Do they have any religions?—Or spiritualism?

The UFO "intervention process" will have a major impact on the US and Russian governments [and] will be the catalyst that will break "the conspiracy of silence." UFOlogists can then divert their focus to the challenges the new alien technologies will bring to mankind.

Due to my declining health, this may be my last writing on this subject. In the meantime, I wish all my readers and UFOlogist supporters the very best in their future work.

STANLEY A. FULHAM, DECEMBER 3RD, 2010[10]

OCTOBER 13, 2010:
DECLOAKING OF AUTHORITY'S SPACECRAFT

The original plan of the regional galactic governance authority, made at its meeting in January 2010 (Earth time), had been for simultaneous decloaking of the authority's spacecraft on October 13, 2010, over major world cities. The purpose of these appearances was to acclimatize

Earth humans to the authority's presence and decision to intervene.

Fulham stated in his Exopolitics radio interview that the regional council had scaled this plan down to one city and had chosen New York City for an initial October 13, 2010, spacecraft "decloaking" because New York City is a global, cosmopolitan city with a blasé population that would not be frightened of their appearance. There are multiple, independent evidentiary sources that prove the October 13, 2010, UFO sightings over New York City revealed intervention by a non-Earthly intelligent civilization and were not the result of other causes, such as released balloons.[11]

Synchronistically, on October 14, 2010, a day after the predicted October 13, 2010, sightings over New York City, Dr. Mazian Othman, director of the UN Outer Space Office, delivered a wide-ranging twenty-eight-minute video press conference at the United Nations in New York, during which she stated that "extraterrestrial life is a possibility" and remarked that the United Nations must ready itself for extraterrestrial contact. Following the October 13, 2010, sightings over New York City, the regional galactic governance authority source predicted a series of sightings of their spacecraft over Moscow, Russia, and London, UK.

WERE THE UFOS IN RUSSIAN SKIES IN DECEMBER 2010 FULHAM'S PREDICTED UFO WAVE?

On December 29, 2010, I reported the following:

> Russian Television (RT) is now reporting UFOs beginning to appear over the skies of Russia and has released a video and news article confirming these sightings. Former NORAD officer Stanley A. Fulham released a statement just prior to his recent death from pancreatic cancer that . . . a regional galactic governance council would begin more frequent UFO appearances in 2011, starting with "a major UFO display over Moscow between the end of the first part of January 2011 and the second week of January 2011."

Query: Is this wave of UFO sightings now beginning over Russia part of the UFO wave predicted by Stanley A. Fulham in his final statement?[12]

One Russian report stated:

In early December UFOs began to frequent the skies of the capital of the republic of Kalmykia, located in southern Russia. Every ten days the residents of Elista have seen UFOs of unknown nature in different parts of the city.

Latest information on this phenomenon appeared on 22 December, 2010, when witnesses observed this type of flying objects in different parts of Elista. So some citizens say they have seen in the sky two concentric circles spinning in opposite directions. Others speak of a triangular object radiating light.

Scientists cannot give a rational explanation for these phenomena. For example, Professor of Theoretical Physics at the University of Kalmykia, Mijaliáyev Badme, does not rule out the possibility of the existence of an alien civilization, but also sees no clear evidence of the existence in the appearance of UFOs in Elista sky. Meanwhile, the ufologists say that UFOs appear in other parts of Russia.[13]

The December 2010 UFO wave phenomenon in Russia does seem related to the extraterrestrial acclimatization process set out in communications from the regional galactic governance council, as published by Stanley A. Fulham in his book *Challenges of Change* and in his December 3, 2010, statement (quoted above in its entirety).[14] Two paragraphs in that statement seem particularly relevant in this regard: "The UFO 'intervention process' will have a major impact on the US and Russian governments. This will be the catalyst that will break the 'conspiracy of silence.'" And "Interventions will then accelerate, not so much over our cities, but dispersed over our continents with sightings increasing in duration. The intent of these interventions is to increase mankind's acceptance of the alien phenomena."

Regarding "the conspiracy of silence" by Russia and the U.S. over the extraterrestrial presence and its intent to save the Earth's ecology—independent analysis has concluded that Fulham's pancreatic cancer may have been accelerated by covert human agencies in order to silence him on the issues of the regional galactic governance council.[15]

Independent Confirmation of Hyperdimensional Council

On January 13, 2011, I reported the following:[16]

> The appearance of [the] apparent Pleiadian UFO craft over Moscow, Russia, marks the second independent occasion on which interdimensional messages inspired by an apparent galactic governance council and predicting an overflight of Pleiadian spacecraft over a major urban area have in fact been accurately fulfilled.
>
> The first occasion was a council prediction through Fulham that galactic governance council ships would appear over New York on the night of October 13, 2010. In fact, UFOs did appear then and were filmed.[17]

The second occasion was the January 2011 UFO sightings over Moscow, Russia. These documented sightings of apparent Pleiadian UFO galactic governance council ships fulfilled Fulham's December 3, 2010, prediction.

LONDON OR PARIS UFO SIGHTINGS IN MID-JANUARY 2011?

According to Fulham's December 3, 2010, communiqué, the next wave of Pleiadian UFO overflight displays would be followed by a major display over London approximately seven days later. A close associate of the now deceased Fulham sent me the following advisory on January 12, 2011.

Stan mentioned he wasn't exactly sure for the longest time if the next event was going to be London or Paris, but after much deliberation, finally settled on London. He was hoping to have another reading with the Transcendors to try to "fine-tune" a more exact date and place, but was too weak and ill. I thought I should mention that to someone in advance, just in case it turns out to be Paris, and the time frame is a bit off.[18]

ASKING THE KEY QUESTION

We need to ask whether the galactic governance council, as described by Fulham and as apparently confirmed by the two independent UFO sightings, might in fact be instead a false flag operation, a psyops, disinformation, or random UFO sightings—or is it what it appears to be, an intervention by a galactic governance authority?

That is a fair question. It has been raised and continues to be raised about the UFO sightings on October 13, 2010, and on November 24, 2010, over New York, and in January 2011 over Moscow. Let us examine the various key aspects of the galactic governance authority intervention hypothesis to date.

Technology of Hyperdimensional ET False Flag Operations

In the absence of any other factors, and considering each of the October, November, and January sightings, a case could be made that each of these may have been either random UFO sightings or a hyperdimensional ET false flag operation, not part of an intentional, interdimensional intervention by an intelligent civilization that so announced itself beforehand. But there are other factors to consider, notably the professional expertise of Fulham regarding extraterrestrial UFOs.

Fulham's Interdimensional Communication

Stanley Fulham was exceptionally capable in recognizing and tracking extraterrestrial UFO craft. As a lifelong NORAD officer, he recounts having routinely tracked extraterrestrial UFOs over North America. From the perspective of a hypothetical regional galactic governance council, Fulham was probably an ideal contact, because he held the highest professional credentials in the recognition of extraterrestrial UFO craft. He spent ten years in study and interdimensional communication around the issue of the regional galactic governance council and its members, functions, and operations. That knowledge was passed onto us in a substantial, serious book.

Timeline of Possible Galactic Governance Council Intervention

The apparent regional galactic governance council's indication of 2014 and 2015 as years for a potential intelligent civilization decloaking an intervention in Earthly affairs is especially significant because of the four (tetra) lunar eclipses separated by two solar eclipses in 2014–15, an astronomical configuration that last occurred in 1492, the year Columbus discovered America, and will not occur again for another 1,000 years.[19]

One interpretation by modern hermeneutics, the analysis of prophecy, views the tetra lunar eclipses of 2014–15 as the prelude to a possible interdimensional landing on Earth of positive intelligent civilizations in a "Messianic" support of human society. This view is congruent with the regional galactic governance council source regarding extraterrestrial intervention in 2014 and 2015, fulfilling a supportive role in the dimensional ecology of Earth.[20]

What if the extraterrestrial interventions predicted by the regional galactic governance council source to occur in 2014 and 2015 do not happen? If they do not, that lack will not detract, in my opinion, from the prevailing evidence suggesting the existence of galactic governance councils, because the promulgation of intentional misinformation or disinformation appears to be a deliberate strategy of these councils when dealing with the hostile environment of terrestrial governments who may want to attack any decloaked spacecraft of the governance council.

EVALUATING THE LEGACY OF STANLEY A. FULHAM

Stanley Fulham's legacy, taken at this most literal level, is potentially of a well-documented revelation of the ecological intervention planned by a regional galactic governance council, and of three separate fulfilled predictions of UFO decloakings of interdimensional craft of the Pleiadians, the reported "point persons" of this council.

The analysis and evaluation by UFO expert Randy Kitchur demonstrates beyond a reasonable doubt that Fulham's predictions of UFO displays over New York City on October 13, 2010, over Moscow early in January 2011, and over London approximately seven days later were literally fulfilled.[21]

Randy Kitchur is the official spokesman for Stan Fulham, author of the book *Challenges of Change,* who accurately predicted UFO displays over New York City on October 13, 2010: "a major UFO display over Moscow between the end of the first part of January 2011 and the second week of January 2011, followed by a major [UFO] display over London approximately seven-days later."

Mr. Kitchur, who spent five years of graduate school with the University of Manitoba department of political studies in Winnipeg, Manitoba, Canada, worked at the University on a master's thesis dealing with the UFO cover-up. He has a year bachelor's of business degree in politics from Brock University in St. Catharines, Ontario, Canada, in addition to two and a half years of sociology/psychology studies at the University of Winnipeg in Winnipeg, Manitoba, Canada.

Stanley A. Fulham and Randy Kitchur made two YouTube videos about Mr. Fulham's book and his October 13th, 2010, UFO predictions in advance of the October 13, 2010 date.

The implications of the fulfillment of the Fulham UFO predictions, when read concurrently with the substance of his well-documented narrative, are substantial. The empirical evidence of the fulfilled Fulham UFO predictions confirms a working hypothesis that a regional galactic governance council, in fact, exists and is now intervening in the human environment for beneficial ecological reasons.

Project Bluebeam is a secret psychological mass deception operation involving three-dimensional optical holograms and sounds, laser projections of multiple holographic images in different parts of the world to create the illusion of mass spacecraft sightings. The Secret Space Program (SSP), financed by what is termed black budget, or secret funding beyond the reach of the constitutional government of the United States, has given rise to substantial fleets of off-planet spaceships and exotic time-space transportation methods such as time travel and teleportation.

Some commentators claim that the UFOs in the New York, Moscow, and London displays are actually part of a Project Bluebeam holographic display, or a display of terrestrial black-budget UFO craft.

This scenario would have it that the Fulham material is actually a covert terrestrial military intelligence "legend" created as part of a false flag operation to impose a global New World Order in order to supposedly defend Earth against an extraterrestrial invasion or landing. A variant of this argument is the claim that the predicted Fulham UFO displays were by an extraterrestrial civilization that is inimical to the aspirations of humanity for sovereignty and self-determination. However, Kitchur's expert UFO evaluation has demonstrated that the interdimensional UFO craft in these three displays are not part of a terrestrial holographic Project Bluebeam operation nor are they terrestrial antigravity UFO craft.[22]

Nor is there any evidence that Fulham was enrolled in a deception operation by a manipulatory extraterrestrial race. Fulham, a deeply spiritual person, as well as a lifelong and highly trained NORAD officer who routinely dealt with UFOs, spent ten years in interdimensional communication around the subject matter of the regional galactic governance council and its message about the ecological crisis of Earth. His writings and actions, as well as the actions of the regional galactic governance council in delivering these materials to him, and in creating the separate interventions over the three cities as evidence of their existence and good faith, are prima facie ethical acts.

At the same time experts in the field of extraterrestrial contactees have commented that contactees of hyperdimensional civilizations are often intentionally given disinformation on certain factual details in order to protect them from attack by human military-intelligence forces on Earth that seek to hide these details. This may well be the case with Stanley Fulham. He appears to have been given accurate information about the dates of the predicted decloakings by the regional galactic governance authority. The rest of the messages of the regional galactic governance authority, about the timing and nature of its intervention in human public affairs, may or may not be intentional disinformation in significant respects. Certainly the predicted dates about the regional galactic governance authority's physical intervention at the United Nations appears to be inaccurate. However, it may or may not be fulfilled at some future date. The issue of the regional galactic governance authority's intervention to help improve Earth's ecology is substantively different, as the regional galactic governance authority could in fact be intervening in the Earth's ecology in ways that are not visible to human observers or may be intervening at some future date. Whatever the authority's strategy it does not appear to have successfully saved Stanley Fulham from assassination, as Mr. Fulham was assassinated by a remotely administered fast-acting cancer, in the opinion of this author, who was one of the last persons to interview Mr. Fulham while he was alive.

PART 3

Spiritual Dimensions of the Omniverse

6

The Interlife Dimensions

Presumptive empirical evidence supports the hypothesis that the Omniverse consists of the totality of parallel universes in the exopolitical dimensions (the multiverse) plus the spiritual dimensions. Evidence for the existence of the spiritual dimensions, for intelligent civilizations of souls* based in the interlife dimensions, and communication between the exopolitical and spiritual dimensions is provided by the same evidence that reveals the continuation of consciousness in the interlife after bodily death. This evidence includes:

- Instrumental transcommunication (ITC) and electronic voice phenomena (EVP)
- Reincarnation studies
- Near-death experiences (NDEs)
- Cross-correspondence experiments through physical mediums

Instrumental Transcommunication (ITC)

Professor at the Technical University (Fachhochschule) Bingen, Ernst Senkowski, Ph.D., of Mainz, Germany, defines instrumental transcommunication (ITC) as Technically supported contacts with "beings,

---

*As stated earlier a *soul* is an individuated, nonlocal, conscious, intelligent entity.

entities, Informational Structures" that are normally not accessible.[1] Using electronic technology, such as video technology, audio technology, computer technology, Internet streaming video or audio, and cell phones,* ITC has provided replicable empirical evidence of interactive communication between living humans in our time-space hologram and the apparent intelligences of deceased humans based in either (a) a hyperdimension of the exopolitical dimension, or (b) the interlife in the spiritual dimensions. Dr. Senkowski states: "Telecommunication devices (audio recorders, radio receivers, telephones, video recorders, TV receivers, computers) behave irregular[ly in] delivering voices, images, and texts of different, sometimes excellent, quality coming from nowhere."[2]

Discussing the data from ITC, researcher Victor Zammit writes:

> According to Mark Macy, a leading researcher in this area, throughout the 1990s, the research laboratories in Europe received extended, two-way communication with spirit colleagues. This was almost daily through telephone answering machines, radios, and computer printouts. His book *Miracles in the Storm* outlines how scientists working for the International Network for Instrumental Transcommunication (INIT), which he founded in 1995, received from the afterlife:

> ❋ pictures of people and places in the afterlife on television that either appeared clearly on the screen and remained for at least several frames, or which built up steadily into a reasonably clear picture over multiple frames;
>
> ❋ text and picture files from people in the afterlife which appeared in computer memory or were planted on disk or similar recordable media;

---

*For sample ITC images from streaming video on Internet and interaction with "deceased" persons in the human "afterlife" dimension (see http://bit.ly/1aFXy9m).

❧ text and images of people and places in the afterlife by way of fax messages.[3]

By producing physical trace evidence in the form of voices and images that correspond to specific identifiable former living and now "dead" persons, ITC provides evidence of both communication between the exopolitical and interlife dimensions as well as survival of the soul after death.

ITC AS NONLOCAL COMMUNICATION

Ervin Laszlo, a noted author, philosopher of science, and systems theorist writes:

The authenticity of ITC is not entirely beyond doubt, but the evidence for it is sufficiently robust to merit sustained investigation. In this writer's view, ITC may be a hitherto unexplored domain of nonlocality, a form of nonlocal communication.

To find a scientific basis for the ITC phenomenon, it must be connected with theories in physics. Transcommunication cannot be connected with classical physics, given that the latter is based on a paradigm that views phenomena not directly traceable to sensory experience as highly suspect, if not clearly illusory. But at the cutting edge of contemporary physics, many things and processes are acknowledged as elements of reality, even when they are intrinsically unobservable. Most pertinently, theories in particle as well as cosmological physics make reference to a field or dimension that subtends the world of the quantum, hitherto considered the lowest level of physical reality. This field or dimensions, variously termed "physical space-time," "nuether," "hyperspace," or "atemporal space" may be responsible for the phenomena of nonlocality in the microdomain of the quantum, as well as in the mesodomain of life and the macrodomain of the universe.[4]

TESLA, MARCONI, EDISON, AND ITC

Professor Ernst Senkowski links the study of ITC to the early beginnings of telecommunications.

Tesla, Marconi, Edison suspected [it might be possible] to contact other worlds by electromagnetic means. Psychic mediums predicted the future development of suitable apparatus. Unexplained signals have been observed since WW I. Partially successful tests were realized in the 20s, 30s, 40s. First voices on tapes were documented independently during the 50s in Italy, USA, Sweden. TV-video images and computer texts appeared later. Official research is not known. Attempts to realize voices on tapes are carried out in at least 12 countries by single operators, small groups, and some bigger associations. Around 60 monographs have been published in seven languages. Only very few operators are scientifically interested. About 10 people have so far realized excellent results.

There are three open questions:

1. How are ITC events realized? Excellent results of these extraordinary psychophysical, mind-matter, man-machine, or consciousness-reality interactions apparently depend upon psychic faculties of the operators. A simple model using the parapsychological terms ESP and PK describes ITC as a largely unconscious two-step process: a trans-entity from beyond our space-time system establishes a telepathic contact with a terrestrial operator who psychokinetically injects the trans-information into the electronic device. It cannot basically be excluded that our technical systems are directly manipulated from the "other side" using techniques unknown to us.

Understanding might result from: consciousness research, neurosciences, psychophysical theories, quantum theoretical models (Bohm, Jahn / Dunne-Pear / Princeton University); sub-quantum or vacuum physics; 12-dimenstional field theory (Burkhard Heim); theory of morphic fields (Sheldrake); system theories; OBE and NDE; thanatology; "far out" speculations [such as] scalar fields and tachyons.

2. What are the contents of ITC? It seems that some "No Body" wants to convince humanity about Life after Death in other states of consciousness and of the existence of nonhuman beings that could be involved to an uncontrollable extent. Generally, the contents confirm the fundamentals of mediumistic channeling in technical forms.

3. Who generates ITC? It seems unjustified to "explain" ITC phenomena as a mere result of the hypothetical "unconscious" of the operator or by super telepathy. Trans-entities appear autonomous and [appear] to be consciously alive. The selectivity of trans-information in cases of drop-in communicators cannot be explained and points to that direction. It is probable that the operator as an active transducer unconsciously adds a certain amount of his own associations to the primary message.

Important issues:

Demonstration of yet undeveloped faculties of the human mind.

Multidimensionality of humans.

Holistic interconnectedness and nonseparability of All-That-Is.

Mental awakening: opening windows to hitherto unknown realities or "parallel worlds," including "past" existences.

An ongoing and future metamorphosis or transformation of humanity.[5]

RELATIONSHIP OF ITC (INTERLIFE) WITH
CHRONOVISION (TIME TRAVEL)

As mentioned earlier it is highly meaningful for the dimensional ecology of the Omniverse hypothesis that ITC's electronic voice phenomena (EVP), used for exploring the interlife in the spiritual dimensions, and chronovision, a time-travel technology for exploring timelines in universes of the exopolitical dimensions, intersect in the work of two Italian Catholic priests, Fathers Pellegrino Ernetti and Augustino Gemelli.

Zammit writes:

[T]he Catholic Church has been actively positive and encouraging towards investigation of the Electronic Voice Phenomena:

Two of the earliest investigators were Italian Catholic priests, Father Ernetti and Father Gemelli, who came upon the phenomena by chance while they were recording Gregorian Chants in 1952. Father Gemelli heard his own father's voice on the tape, calling him by a childhood nickname, saying: "Zucchini, it is clear, don't you know it is I?"[6]

Deeply troubled by Catholic teaching in regard to contact with the dead, the two priests visited Pope Pius XII in Rome. The Pope reassured them: "Dear Father Gemelli, you really need not worry about this. The existence of this voice is strictly a scientific fact and has nothing to do with spiritism. The recorder is totally objective. It receives and records only sound waves from wherever they come. This experiment may perhaps become the cornerstone for a building for scientific studies which will strengthen people's faith in a hereafter."[7]

Pope Pius's cousin, the Rev. Professor Gebhard Frei, Ph.D., cofounder of the Jung Institute, was an internationally known parapsychologist who worked closely with Raudive, a pioneer in the research. He was also the president of the International Society for Catholic

Parapsychologists. He himself is on record as stating, "All that I have read and heard forces me to believe that the voices come from transcendental, individual entities. Whether it suits me or not, I have no right to doubt the reality of the voices."[8]

Frei died on 27 October, 1967. In November 1967, at numerous taping sessions, a voice giving its name as Gebhard Frei came through. The voice was identified by Professor Peter Hohenwarter of the University of Vienna as positively belonging to Frei.[9]

Pope Paul VI was well aware of the work being done from 1959 onward on the electronic voices by his good friend, Swedish film producer Friedrich Jürgenson, who had made a documentary film about him. The Pope made Jürgenson a Knight Commander of the Order of St. Gregory in 1969 for his work. Jürgenson wrote to Bander, "I have found a sympathetic ear for the Voice Phenomenon in the Vatican. I have won many wonderful friends among the leading figures in the Holy City. Today 'the bridge' stands firmly on its foundations."[10]

Former U.S. chrononaut Andrew D. Basiago notes that Father Pellegrino Ernetti, discoverer of chronovision, states, "From birth to death, every individual traces an arc of light and sound in the quantum hologram."*[11]

By the "quantum hologram," Basiago and Father Pellegrino Ernetti are referring to dimensions in the universes of the exopolitical dimension of the Omniverse.

The story of Ernetti's original discovery of the instrumental transcommunication (ITC) capabilities of his machine through interactive communication with his deceased father, and a translation of a 1972 Italian newspaper article, "Invention of a machine for photographing

---

*According to former U.S. chrononaut Andrew D. Basiago, "The chronovisor was accidentally discovered when the Benedictine priests Pellegrino Ernetti and Augustino Gemelli were investigating the harmonic patterns in Gregorian chants at the Catholic University of Milan in the 1940's."

the past," establish at least two applications of the ITC/chronovisor as developed by Ernetti and Gemelli. The chronovisor of Ernetti and Gemelli was used to achieve:

✸ Instrumental transcommunication (ITC) resulting in interactive communication with the intelligences of specific deceased humans located in the interlife in the spiritual dimensions of the Omniverse

✸ Chronovision time travel: Location and video and audio recording of events along a past timeline in the time-space hologram of the multiverse

The existence of time travel along timelines in the time-space hologram (quantum hologram) in the multiverse can be established through Ernetti and Gemelli's reported chronovision recordings of famous personages in history. Ernetti himself states that the chronovisor can tune in to individual voices and images along a timeline in the exopolitical dimension,

with the use of suitable equipment, [which] our team was the first in the world to build. The equipment consists of a series of antennas to allow tuning of the individual voices and images. We know that every human being, from when he was born to when he dies, leaves behind himself a double wake: a visual and a sound, a sort of identity card for each different person. And based on this different identity card one can reconstruct the single person in all his deeds and his sayings, and for this reason we are now able to hear and see again the greatest people in history.[12]

Ernetti and Gemelli's experiments provide empirical evidence supporting: the dimensional ecology hypothesis of an Omniverse that consists of both exopolitical and spiritual dimensions; that we humans exist in a time-space hologram as incarnations in the exopolitical dimensions; and that souls are based in the interlife in the spiritual dimensions.

Reincarnation Studies

Prevailing empirical evidence for the dimensional ecology hypothesis and for existence of the intelligent civilizations of souls based in the interlife of the spiritual dimensions has been developed through science-based studies suggestive of reincarnation. Reincarnation is a fundamental process in the Omniverse. By means of it a soul based in the interlife dimensions of the spiritual dimensions can transition into a parallel dimension or universe of the multiverse and can undertake an incarnated life, using the support system of the physical body of some creature or entity as an avatar, usually in an intelligent civilization. Some spiritual traditions speak of the transmigration of human souls into the animal, vegetable, or mineral kingdoms as well. Our focus here is on the study of reincarnation of souls as intelligent civilizations.

The two principal methods of scientific scrutiny of reincarnation in the exopolitical dimensions are hypnotic regression to capture memories of prior (and future) lives, and study of spontaneous memories of past lives among children. We will focus on hypnotic regression in chapter 7 in reconstructing the nature of the afterlife or interlife dimension.

SPONTANEOUS MEMORIES OF PAST LIVES

Dr. Ian Stevenson (1918–2007), formerly of the University of Virginia Medical School, studied more than 3,000 cases suggestive of reincarnation in children who had spontaneous memories of apparent prior lives. These cases provide proof of the continuation of consciousness after physical death, because each case provides empirically verified evidence of the continuation of an underlying consciousness (soul) in a new physical body.*

There are common patterns in all these cases, in that the children who had spontaneous memories of prior lives began relating facts and stories about their prior lives between the ages of two and four years old. By the age of eight or nine years old, the children by and large appeared to have forgotten about the prior lives. Other patterns in these cases are the frequent mention by the children of having died a violent death in the prior life and a clear memory of the mode of death.

Memories of a Past Life

A summary of Dr. Ian Stevenson's cases demonstrates patterns of evidence at first sight that support the dimensional ecology hypothesis. In most of Dr. Stevenson's cases, the same underlying soul is demonstrated as carrying memories from a prior incarnation as an Earthling human in the time-space hologram of the exopolitical dimensions. One summary states:

> The most frequently occurring event or common denominator relating to rebirth is probably that of a child remembering a past life. Children usually begin to talk about their memories between the ages of two and four. Such infantile memories gradually dwindle when the child is between four and seven years old. There are, of course, always some exceptions, such as a child continuing to remember its previous life but not speaking about it for various reasons.
>
> Most of the children talk about their previous identity with great intensity and feeling. Often they cannot decide for themselves which world is real and which one is not. They often experience a kind of double existence, where at times one life is more prominent,

---

*For an overview of Stevenson's work, see University of Virginia Medical School, The Division of Perceptual Studies, http://bit.ly/17cJkMo (accessed June 30, 2015).

and at times the other life takes over. This is why they usually speak of their past life in the present tense, saying things like, "I have a husband and two children who live in Jaipur." Almost all of them are able to tell us about the events leading up to their death.

Such children tend to consider their previous parents to be their real parents rather than their present ones, and usually express a wish to return to them. When the previous family has been found and details about the person in that past life have come to light, then the origin of the fifth common denominator—the conspicuous or unusual behavior of the child—is becoming obvious.

For instance, if the child is born in India to a very low-class family and was a member of a higher caste in its previous life, it may feel uncomfortable in its new family. The child may ask to be served or waited on hand and foot and may refuse to wear cheap clothes. Stevenson gives us several examples of these unusual behavior patterns.[13]

Birthmarks and Wounds Suffered in a Prior Life

The dimensional ecology hypothesis is supported by the rigor of Dr. Ian Stevenson's cases that illustrate how energetic impressions may persist in the dimensional ecology over multiple incarnations of a particular soul, depending on the soul's karmic attachment to events from prior lifetimes. For example, another empirical regularity Stevenson demonstrates, one that is presumptive evidence, is that a specific soul may manifest wounds suffered in a prior life as birthmarks in a present life in the Earth's time-space hologram in the exopolitical dimensions. Stevenson writes:

Almost nothing is known about why pigmented birthmarks (moles or nevi) occur in particular locations of the skin. The causes of most birth defects are also unknown. About 35 percent of children who claim to remember previous lives have birthmarks and/or birth defects that they (or adult informants) attribute to wounds on a person whose life the child remembers. The cases of 210 such children have been investigated. The birthmarks were usually areas of hairless, puckered skin; some were areas of little or no pigmentation (hypopigmented macules); others were areas of increased pigmentation (hyperpigmented nevi). The birth defects were nearly always of rare types. In cases in which a deceased person was identified, the details of whose life unmistakably matched the child's statements, a close correspondence was nearly always found between the birthmarks and/or birth defects on the child and the wounds on the deceased person. In 43 of 49 cases in which a medical document (usually a postmortem report) was obtained, it confirmed the correspondence between wounds and birthmarks (or birth defects). There is little evidence that parents and other informants imposed a false identity on the child in order to explain the child's birthmark or birth defect. Some paranormal process seems required to account for at least some of the details of these cases, including the birthmarks and birth defects.

Because most (but not all) of these cases develop among persons who believe in reincarnation, we should expect that the informants for the cases would interpret them as examples according with their belief; and they usually do. It is necessary, however, for scientists to think of alternative explanations.

The most obvious explanation of these cases attributes the birthmark or birth defect on the child to chance, and the reports of the child's statements and unusual behavior then become a parental fiction intended to account for the birthmark (or birth defect) in terms of the culturally accepted belief in reincarnation. There are, however, important objections to this explanation.

First, the parents (and other adults concerned in a case) have no need to invent and narrate details of a previous life in order to explain their child's lesion. Believing in reincarnation, as most of them do, they are nearly always content to attribute the lesion to some event of a previous life without searching for a particular life with matching details.

Second, the lives of the deceased persons figuring in the cases were of uneven quality both as to social status and commendable conduct. A few of them provided models of heroism or some other enviable quality; but many of them lived in poverty or were otherwise unexemplary. Few parents would impose identification with such persons on their children.

Third, although in most cases the two families concerned were acquainted (or even related), I am confident that in at least 13 cases (among 210 carefully examined with regard to this matter) the two families concerned had never even heard about each other before the case developed. The subject's family in these cases can have had no information with which to build up an imaginary previous life that, it later turned out, closely matched a real one. In another 12 cases the child's parents had heard about the death of the person concerned, but had no knowledge of the wounds on that person. . . .

Fourth, I think I have shown that chance is an improbable interpretation for the correspondences in location between two or more birthmarks on the subject of a case and wounds on a deceased person.[14]

Stevenson places his research in the context of reasoned consideration of alternative explanations:

Persons who reject the explanation of chance combined with a secondarily confected history may consider other interpretations that include paranormal processes, but fall short of proposing a life after death. One of these supposes that the birthmark or birth defect occurs by chance and the subject then by telepathy learns about a deceased person who had a similar lesion and develops identification with that person. The children subjects of these cases, however, never show paranormal powers of the magnitude required to explain the apparent memories in contexts outside of their seeming memories.

Another explanation, which would leave less to chance in the production of the child's lesion, attributes it to a maternal impression on the part of the child's mother. According to this idea, a pregnant woman, having knowledge of the deceased person's wounds, might influence a gestating embryo and fetus so that its form corresponded to the wounds on the deceased person. The idea of maternal impressions, popular in preceding centuries and up to the first decades of this one, has fallen into disrepute. Until my own recent article (Stevenson, 1992) there had been no review of [the] series of cases since 1890 . . . Nevertheless, some of the published cases—old and new—show a remarkable correspondence between an unusual stimulus in the mind of a pregnant woman and an unusual birthmark or birth defect in her later-born child. Also, in an analysis of 113 published cases I found that the stimulus occurred to the mother in the first trimester in 80 cases (Stevenson, 1992). The first trimester is well known to be the one of greatest sensitivity of the embryo/fetus to recognized teratogens, such as thalidomide. . . . Applied to the present cases, however, the theory of maternal impression has obstacles as great as the normal explanation appears to have. First, in the 25 cases mentioned above, the subject's mother, although she may have heard of the death of the concerned deceased person, had no knowledge of that person's wounds. Second, this interpretation supposes that the mother not only modified the body of her unborn child with her thoughts, but also after the child's birth influenced it to make statements and show behavior that it otherwise would not have done. No motive for such conduct can be discerned in most of the mothers (or fathers) of these subjects.[15]

Dr. Ian Stevenson's cases thus provide prima facie and probable cause evidence for reincarnation as a regularity of the dimensional ecology, as it is most probably the reincarnating soul's wounds in a prior life that are manifesting as birthmarks in a present life due to an unresolved energetic or karmic impression on behalf of the reincarnating soul. As a whole Dr. Stevenson's cases also provide support for the thesis that the mission of the universes of the multiverse is to serve as venues for reincarnation by souls for moral and soul development.

Near Death Experiences (NDEs)

Other methodologies exist for developing replicable empirical evidence supporting the dimensional ecology hypothesis of the Omniverse. For example, David Fontana, psychologist, parapsychologist, author, and former professor of psychology at Cardiff University, lists a total of seventeen formal methodologies for proving the existence of the interlife dimensions.[16] Notable among these methodologies is the study of near death experiences (NDEs), which provide presumptive eyewitness evidence of the transition of the human soul through the dimensions and of the spiritual dimension itself.[17]

Victor Zammit, an Australian lawyer who has become a specialist on the evidence for life after death, has summarized fourteen points to show why NDEs are real and not hallucinations produced by lack of brain oxygen or drugs. He writes:

1. NDE survivors have clear and structured memories of what happened to them. Patients who did not have a NDE during similar treatment were confused or could not remember anything.

2. Whereas hallucinations are all different, near-death experiences are very similar in different cultures and throughout history. Near-death experiences have been reported in all cultures, and from as far back as 1760 BC.

3. People see and hear things while they are unconscious that

would be impossible for normal sensing. A huge percentage of near-death experiencers are able to describe exactly what happened to them while they were unconscious.

4. People come back from a NDE with accurate facts they did not know before.

5. People report meeting with relatives they did not know were dead. In all cases, they are correct.

6. Some people come back with knowledge of the future.

7. Some people come back with advanced knowledge consistent with quantum physics. Almost all survivors say that they entered a dimension where there was no time and many were able to go back and forward through time.

8. Some people are cured of fatal illnesses during a NDE or have miraculous recoveries from serious injuries.

9. The blind can see during a NDE.

10. Some people have a group near-death experience. A group of fire fighters claimed that when they were overcome with smoke, they all went out of their bodies. They communicated with each other and could all see the lifeless bodies below them.

11. Some people have near-death-like experiences when there is nothing physically wrong with them. Researchers have found that deep meditation, deathbed visions, relaxation, psychic vision, astral projection, trance, mirror-gazing, and eye movements can trigger elements of the NDE.

12. Some people have a near-death experience when they are completely brain dead. Hallucinations can only occur when people have a functioning brain which shows an active EEG reading.

13. Many people experience a "life review" during which they see their lives from the perspective of other people.

14. The after-effects of a NDE are unique and long lasting.

There have been many attempts to explain away the near-death experience. Some claim they are caused by oxygen deprivation. Others claim it is a natural effect of the dying brain or some accident of brain chemistry. Most of these theories are based on observations of a small number of cases. They may produce elements of the near-death experience but not the whole experience. And, most important, they do not have the same impact or after-effects.[18]

As an attorney trained for twenty-five years in the law of evidence, Victor Zammit has been able to assemble a critical mass of replicable, empirical evidence on near death experiences (NDEs), giving rise to probable cause of the continuation of the life of the soul after bodily death in the specific dimensions of the afterlife, thus supporting crucial elements of the dimensional ecology of the Omniverse hypothesis.

Cross-Correspondence Experiments through Physical Mediums

Another notable methodology is controlled experiments involving physical mediumship. One experiment using physical mediums to develop evidence of the human soul's survival in the interlife in the spiritual dimensions was that of cross-correspondences, designed by Cambridge classics scholar Frederick W. H. Myers, who—after he died—transmitted a series of signed scripts through more than a dozen mediums in different parts of the world.

Zammit writes:

Later, there were scripts signed by his fellow leaders of the Society for Psychical Research, Professor Henry Sidgwick and Edmund Gurney, as they too had died. The scripts were all about unusual classical subjects and did not make sense on their own. But the mediums were told to contact a central address and send the scripts there.

When the scripts were put together they fitted like pieces of a jig-saw puzzle. In all, more than 3,000 scripts were transmitted over 30 years. Some of them were more than 40 typed pages long. Together they fill 24 volumes and 12,000 pages. The investigation went on so long that some of the investigators, such as Professor Verrall, died during the course of it and began communicating themselves.[19]

Likewise, the experimental results of cross correspondences support the afterlife hypothesis and, in so doing, in turn support crucial elements of the dimensional ecology of the Omniverse hypothesis by demonstrating how an intelligence can function in the interlife and coordinate activities in the time-space world of Earth.

Biocentrism and the Dimensional Ecology Hypothesis

Biocentrism is a new school of thought based on Robert Lanza's book *Biocentrism* that posits that life and consciousness, rather than life-less processes of physics, are central to the origins and processes of our Universe.

Even scientist Lanza's biocentrism is supportive of the perspective of the Omniverse presented in the dimensional ecology hypothesis.

There is no separate physical universe outside of life and conscious-ness. Nothing is real that is not perceived. There was never a time when an external, dumb, physical universe existed, or that life sprang randomly from it at a later date. Space and time exist only as constructs of the mind, as tools of perception. Experiments in which the observer influences the outcome are easily explainable by the interrelatedness of consciousness and the physical universe. Neither nature nor mind is unreal; both are correlative. No position is taken regarding God.[20]

One analysis of biocentrism by writer Anna Lemind concludes:

The theory [Biocentrism] implies that death simply does not exist. It is an illusion which arises in the minds of people. It exists because people identify themselves with their body. They believe that the body is going to perish, sooner or later, thinking their consciousness will disappear too. In fact, consciousness exists outside of constraints of time and space. It is able to be anywhere: in the human body and outside of it. That fits well with the basic postulates of quantum mechanics science, according to which a certain particle can be present anywhere and an event can happen according to several, sometimes countless, ways.

Lanza believes that multiple universes can exist simultaneously. These universes contain multiple ways for possible scenarios to occur. In one universe, the body can be dead. And in another it continues to exist, absorbing consciousness which migrated into this universe.

This means that a dead person while traveling through the same tunnel ends up not in hell or in heaven, but in a similar world he or she once inhabited, but this time alive. And so on, infinitely. . . .

So, there is an abundance of places or other universes where our soul could migrate after death, according to the theory of neo-biocentrism. But does the soul exist?

Professor Stuart Hameroff from the University of Arizona has no doubts about the existence of [an] eternal soul. As recently as last year, he announced that he has found evidence that consciousness does not perish after death.

According to Hameroff, the human brain is the perfect quantum computer and the soul or consciousness is simply information stored at the quantum level. It can be transferred, following the death of the body; quantum information represented by consciousness merges with our universe and exists there indefinitely. The biocentrism expert Lanza proves that the soul migrates to another universe. That is the main difference from his other colleagues.

Sir Roger Penrose, a famous British physicist and expert in mathematics from Oxford, supports this theory, and he has also found traces of contact with other universes. Together, the scientists are developing quantum theory to explain the phenomenon of consciousness. They believe that they found carriers of consciousness, the elements that accumulate information during life and [that], after death of the body . . . "drain" consciousness somewhere else. These elements are located inside protein-based microtubules (neuronal microtubules), which previously have been attributed a simple role of reinforcement and transport channeling inside a living cell. Based on their structure, microtubules are best suited to function as carriers of quantum properties inside the brain . . . because they are able to retain quantum states for a long time [and can thus] function as elements of a quantum computer.[21]

In many ways Robert Lanza and biocentrism have developed a parallel proof of the dimensional ecology of the Omniverse hypothesis, as Lanza has arrived at many of the same basic tenets without explicitly investigating the same replicable empirical data explored in this book.

The centrality of consciousness forms an essential backdrop to understanding the intelligent civilizations of souls, examined in detail in the next chapter.

7

The Intelligent Civilizations of Souls

Congruent with the dimensional ecology of the Omniverse hypothesis, the life of souls has long been recorded in human history. One ancient example is provided by Plato's descriptions of the soul, as recounted by Anthony Peake, an author and researcher on the human afterlife.

> Plato is clear that the soul survives after death. He describes, through the experiences of Er, that some form of judgment takes place. This is made by celestial beings that, although it is not stated explicitly, seem to review the life of the soul in question. A decision is then made as to whether the soul is reborn again ("to the left and downward") or is allowed to move on to a higher plane of existence. This image we will see described again and again.[1]

The dimensional ecology of the Omniverse hypothesis is also supported by replicable data from hypnotic regression. During 1960 to 1980 Michael Newton developed and applied standard protocols in using hypnotic regression to access soul memories. His data eventually grew to more than 7,000 replicable cases, in a body of work that was initially published in 1995, more than thirty years after it was commenced.[2] This presumptive empirical evidence gives us detailed information about the interactions of the intelligent civilizations of souls, of spiritual beings, and of Source (God) with the universes and intelligent civilizations of the exopolitical dimensions (multiverse).

Specifically, our review of representative excerpts and summaries by Michael Newton and his associates has found supportive evidence for:

⚜ Nature and creation of souls as holographic fragments of God

⚜ Soul development through all the dimensions of the Omniverse (and the role of soul groups)

⚜ Reentry from the exopolitical dimensions into the interlife dimensions

⚜ Time travel, teleportation, and interdimensional portals between exopolitical and interlife dimensions

⚜ Life creation in the universes of the exopolitical dimensions (multiverse)

⚜ Nature of Source (God) and relations between exopolitical and spiritual dimensions

⚜ Soul creation of galactic matter, stars, and planets in the exopolitical dimensions (including interdimensional travel and alterations to time and matter)

⚜ Soul interventions in dreams of individuals in the universes of the exopolitical dimensions

⚜ Role of Grey extraterrestrials in soul's reincarnation cycle

⚜ Parallels between hyperdimensional extraterrestrials and souls

We encourage you to read all of Michael Newton's original research (listed in the bibliography) to experience the full import of his discoveries about the nature of all of the dimensions of the Omniverse.

Nature and Creation of Souls
as Holographic Fragments of God

Newton defines the soul as "intelligent light energy." He writes:

> We cannot define the soul in a physical way because to do so would
> establish limits on something that seems to have none. I see the soul
> as intelligent light energy. This energy appears to function as vibra-
> tional waves similar to electromagnetic force but without the limita-
> tions of charged particles of matter. Soul energy does not appear to
> be uniform. Like a fingerprint, each soul has a unique identity in its
> formation, composition, and vibrational distribution. I am able to
> discern soul properties of development by color tones, yet none of
> this defines what the soul is as an entity.[3]

One conclusion supported by Newton's evidence regards the nature
and origin of souls: each one is a holographic fragment of the original Source or Creator: they are holographic "eggs of Light" from God,
who manifests as a "Sea of Light." In the following recorded regression,
Newton is speaking with a soul that works at a soul-creation facility in
the spiritual dimensions.

NEWTON: As an Incubator Mother, when do you first see the new
souls?

SOUL: We are in the delivery suite, which is a part of the nursery, at
one end of the emporium. The newly arrived ones are conveyed
as small masses of white energy encased in a gold sac. They move
slowly in a majestic, orchestrated line of progression toward us.

NEWTON: From where?

SOUL: At our end of the emporium under an archway the entire wall is
filled with a molten mass of high-intensity energy and . . . vitality. It

feels as if it's energized by an amazing love force rather than a discernible heat source. The mass pulsates and undulates in a beautiful flowing motion. Its color is like that on the inside of your eyelids if you were to look through closed eyes at the sun on a bright day.

NEWTON: And from out of this mass you see souls emerge?

SOUL: From the mass a swelling begins, never exactly from the same site twice. The swelling increases and pushes outward, becoming a formless bulge. The separation is a wondrous moment. A new soul is born. It's totally alive with an energy and distinctness of its own.[4]

Newton added the following note about incubation from another interview with a high-level soul.

I see an egg-shaped mass with energy flowing out and back in. When it expands, new soul energy fragments are spawned. When the bulge contracts, I think it pulls back those souls which were not successfully spawned. For some reason these fragments could not make it on to the next step of individuality.[5]

Soul Development through All the Dimensions of the Omniverse

Another conclusion supported by Newton's evidence is that the intelligent civilizations of souls based in the spiritual dimensions of the Omniverse incarnate in the multiverse to undergo a variety of life experiences as a diversity of intelligent creatures for the purpose of individual moral and soul development. As the soul is a holographic fragment of Source (God), each soul's individual development advances the development of the collective spiritual dimension.

The universes of the exopolitical dimensions function as a virtual experiential development environment for souls. The major process of soul development takes place through incarnation in a universe of the exopolitical dimensions as an intelligent species. One function of the local dimensions of consciousness in our universe is to provide stratification and diversity in the evolution of consciousness, as well as to promote the evolution of consciousness across dimensions.

For example, as mentioned, Earth humans are based in the third time-space dimension. More evolved humans, such as Arcturian humans, have described themselves as based in the sixth light dimension of our universe, according to one system of consciousness stratification. In terms of their relative position in our universe's dimensions of consciousness, Arcturian humans are more evolved than Earth humans. Souls choosing to incarnate from the interlife would experience vastly different lives in the exopolitical dimensions as Arcturian humans than they would experience as Earth humans. The soul development for a soul incarnating as an Earth human could, in theory, be greater in magnitude than that experienced by a soul incarnating as an Arcturian.[6]

The specific steps in the incarnation process include:

1. Selecting a future physical life, including technological "sampling of lives"
2. Rehearsal and recognition classes for chosen future lives with incarnating soul mates
3. Teleportation (in a normal incarnation and birth) through an interdimensional portal from the interlife dimensions into the exopolitical dimensions to coordinates next to the womb of the prospective mother, enmeshing into the fetus of the soul's new life about three months prior to birth (final trimester)

In unusual cases a single soul may lead simultaneous, different physical lives in different locations of the exopolitical dimensions. In normal cases, at physical death, the soul teleports back through an interdimensional portal into the interlife and (unless an advanced soul) is met by guides and family, friends, and so on. A life-review period follows with guides and a council of elders, including a holographic review of life episodes for karmic lessons, and so on. The soul enters a transition period for further education, new soul career as guide, healer, new life planning, and so on. Then the soul is placed in this new area of activity, and where necessary a new life-selection process begins.

SOUL GROUPS IN THE INTERLIFE DIMENSIONS

Data from hypnotic regression indicates that in the interlife soul development and education takes place through groups. Souls are assigned to soul groups with which they tend to remain, in a context of total freedom. In reincarnation soul groups may tend to navigate the dimensions together, incarnated as groups in varying social roles on planets in the universes of the exopolitical dimensions. Newton's data indicates:

1. New souls are assigned to a new soul support group based on their level of understanding.
2. Once a new soul support group is formed, new soul members are not added in the future.
3. There is a process for systematic selection of homogeneous groups of souls. Similarities in ego, cognitive consciousness, expression, and desire are defining considerations.
4. Regardless of size soul groups do not mix their energy with that of other soul groups. Souls can communicate between themselves.
5. Learning curves can differ for individual souls. Some souls advance more rapidly than others. At the intermediate level, should they show special talents (healing, creativity, teaching, etc.), they are authorized to participate in special training while still in their soul groups.

6. At the intermediate level souls may join together in large independent study groups.

7. At the more advanced level souls can undertake independent activities outside the group and come to act as guides.

8. A classification model for soul development has found the following mix of soul development levels and auras:

>Beginner's souls, 42 percent (white aura)
>
>Intermediate souls, 57 percent (yellow-gold aura)
>
>Advanced, elevated souls, 1 percent (light blue-purple aura)[7]

Reentry from the Exopolitical Dimensions into the Interlife Dimensions

The data from hypnotic regression of soul memories of the interlife indicate that there are standard treatments for souls reentering the spiritual interlife dimensions via the interdimensional portal. The soul guides use three techniques at the portal:

1. Envelopment. Entering souls are enveloped in a large circular mass of energy by their guide to ease the transition from incarnation in the exopolitical dimension.

2. Focus effect. The guide can apply energy to border points, as a sort of spiritual acupuncture, along meridians of the soul's etheric body.

3. Emergency treatment at the interdimensional portal. If an arriving soul has experienced the violent death of the physical body it was incarnated in, the soul may arrive with its energy in a deteriorated state. The guides will then give energetic and spiritual mediation exercises to the soul before it moves forward into the interlife and the spiritual dimensions.[8]

Time Travel, Teleportation, and
Interdimensional Portals

Probable empirical evidence exists, supporting the dimensional ecology hypothesis, that souls from the interlife in the spiritual dimensions incarnate (by teleportation) as physical lives in other universes of the exopolitical dimensions. Newton states, "I have an extensive file on souls that have incarnated in other worlds and souls that have traveled in a variety of unknown worlds for study and recreation between their lives on Earth."[9]

The replicable data from hypnotic regression reported by Newton and his associates are congruent with a report made to me by Andrew D. Basiago, about having seen and heard recently deceased souls while time-travel teleporting as a U.S. chrononaut. These souls were in dimensions or "astral planes" adjacent to the vortal tunnel that Basiago was teleporting through. Apparently, these souls were still in transition in the time-space hologram of our universe, on their way to the interdimensional portal to the interlife dimensions.

Reporting on descriptions of this transition, Newton writes:

Once through the tunnel, our souls have passed the initial gateway of their journey into the spirit world. Most now fully realize they are not really dead, but have simply left the encumbrance of an Earth body which has died. With this awareness comes acceptance in varying degrees depending upon the soul. Some subjects look at these surroundings with continued amazement while others are more matter-of-fact in reporting to me what they see. Much depends upon their respective maturity and recent life experiences. The most common type of reaction I hear is a relieved sigh followed by something on the order of, "Oh, wonderful, I'm home in this beautiful place again." There are those highly developed souls who move so fast out of their bodies that much of what I am describing here is a blur as they home in to their spiritual destinations. These are the pros and, in my opinion, they are a distinct minority on Earth. The average soul does not move that rapidly and some are very hesitant. If we exclude the rare cases of highly disturbed spirits who fight to stay connected with their dead bodies, I find it is the younger souls with fewer pasts.

Most of my subjects report that as they emerge from the mouth of the tunnel, things are still unclear for a while. I think this is due to the density of the nearest astral plane surrounding Earth, called the *kamaloka* by Theosophists.[10]

Life Creation in the Universes of the Exopolitical Dimensions

Replicable data developed by Newton and his associates illustrate the roles that the intelligent civilizations of souls have played in creating life in the multiverse. These roles are consistent with the replicable finding that "God" is the totality of the spiritual dimension of the Omniverse, namely, the intelligent civilizations of souls, of spiritual beings, and God itself.

In this transcript Newton illustrates how a soul can participate in creating life:

NEWTON: Let me try and sum this up, and please tell me if I am on the wrong track. A soul who becomes proficient with actually creating life must be able to split cells and give DNA instructions, and you do this by sending particles of energy into protoplasm?

SOUL: We must learn to do this, yes, coordinating it with a sun's energy.

NEWTON: Why?

SOUL: Because each sun has different energy effects on the worlds around them.

NEWTON: Then why would you interfere with what a sun would naturally do with its own energy on a planet?

SOUL: It is not interference. We examine new structures . . . mutations . . . to watch and see what is workable. We arrange substances for their most effective use with different suns.

NEWTON: When a species of life evolves on a planet, are the environmental conditions for selection and adaptation natural, or are intelligent soul-minds tinkering with what happens?

SOUL: (*evasively*) Usually a planet hospitable to life has souls watching and whatever we do is natural.

NEWTON: How can souls watch and influence biological properties of growth evolving over millions of years on a primordial world?

SOUL: Time is not in Earth years for us. We use it to suit our experiments.

NEWTON: Do you personally create suns in our universe?

SOUL: A full-scale sun? Oh no, that's way over my head . . . and requires the powers of many. I generate only on a small scale.

NEWTON: What can you generate?

SOUL: Ah . . . small bundles of highly concentrated matter . . . heated.

NEWTON: But what does your work look like when you are finished?

SOUL: Small solar systems.

NEWTON: Are your miniature suns and planets the size of rocks, buildings, the moon—what are we talking about here?

SOUL: (laughs) My suns are the size of basketballs and the planets . . . marbles . . . that's the best I can do.

NEWTON: Why do you do this on a small scale?

SOUL: For practice, so I can make larger suns. After enough compression the atoms explode and condense, but I can't do anything really big alone.

NEWTON: What do you mean?

SOUL: We must learn to work together to combine our energy for the best results.

NEWTON: Well, who does the full-sized thermonuclear explosions which create physical universes and space itself?

SOUL: The Source . . . the concentrated energy of the Old Ones.

NEWTON: Oh, so the Source has help?[11]

Nature of Source (God) and Relations between Exopolitical and Spiritual Dimensions

Newton's data support the dimensional ecology of the Omniverse hypothesis by describing aspects of a spiritual dimension that: (1) consists of the intelligent civilizations of souls, of spiritual beings, and of God itself; and (2) functionally is the Source of the created universes of time, space, energy, and matter in the exopolitical dimensions (multiverse).

NEWTON: Tell me, does the Source dwell in some special central space in the spirit world [spiritual dimensions]?

SOUL: The Source is the spirit world [spiritual dimensions].[12]

NEWTON: If the Source represents the entire spirit world, how does this mental place differ from physical universes with stars, planets, and living things?

SOUL: Universes are created—to live and die—for the use of the Source. The place of spirits . . . is the Source.

NEWTON: We seem to live in a universe which is expanding and may contract again and eventually die. Since we live in a space with time limitations, how can the spirit world itself be timeless?

SOUL: Because here we live in non-space which is timeless . . . except in certain zones.

NEWTON: Please explain what these zones are.

SOUL: They are . . . interconnecting doors . . . openings for us to pass through into a physical universe of time.

NEWTON: You speak of universes in the plural. Are these other physical universes besides the one which contains Earth?

SOUL: (*vaguely*) There are . . . differing realities to suit the Source.

NEWTON: Are you saying souls can enter various rooms of different physical realities from spiritual doorways?

SOUL: (*nods*) Yes, they can—and do.[13]

Soul Creation and Adjustment of Galactic Matter, Stars, and Planets in the Exopolitical Dimensions

Newton's persuasive evidence shows the extensive role that the intelligent civilizations of souls have in both the spiritual dimensions and in the universes of the exopolitical dimensions (multiverse). Souls have an extensive role in the planning, creation, and husbandry of universes, galaxies, planets, and celestial bodies, as well as of living species of flora and fauna in the multiverse. In regression interviews with an advanced soul of the "Structural Soul" category, Newton was able to establish the role of advanced souls in creating galactic matter, stars, and planets in the universes of the exopolitical dimensions. This replicable data further supports the dimensional ecology of the Omniverse hypothesis. Newton writes:

> My subjects who are Structural Souls say these designs relate to the formation of "geometric shapes that float as elastic patterns," which contribute to the building blocks of a living universe. . . . The Masters of Design have enormous influence on creation. I'm told they are capable of bridging universes that seem not to have a beginning or end, exacting their purposes among countless environmental settings. Carried to its logical conclusion, this would mean these masters—or grandmasters—would be capable of creating the spinning gas clouds of galactic matter which started the process of stars, planets.[14]

His data also reveal that advanced souls of the "Explorer" category travel from the interlife dimensions to the universes of the exopolitical dimensions (multiverse) while they are creating or adjusting planets. Newton writes:

> Souls who travel interdimensionally explain that their movements appear to be in and out of curved spheres connected by zones that are opened and closed by converging vibrational attunement. Explorer [soul] trainees have to learn this skill. From the accounts I have heard, the interdimensional [soul] travelers must also learn about the surface boundaries of zones as hikers locating trailheads between mountain ranges. Souls speak of points, lines, and surfaces in multi-space which indicate larger structural solids, at least for the physical universes. I would think dimensions having geometric designs need hyperspace to hold them. Yet Explorer souls travel so fast in some sort of hyperspace it seems to me the essence of speed, time, and direction of travel is hardly definitive. Training to be an Explorer [soul] must indeed be formidable, as indicated by a

quote from this client [whose soul] travels through five dimensions between her lives:

These dimensions are so enmeshed with one another that I have no sense of boundaries except for two elements, sound and color. I must learn to attune my energy to the vibrational frequency of each dimension, and some are so complex I cannot yet go to them. With color, the purples, blues, yellows, reds and whites are manifestations of light and density for those energy particles in the dimensions where I travel.[15]

Data from hypnotic regression also reveal incarnations on simulated planets in the exopolitical dimensions for purposes of soul development of specific skills, such as mastery of the time dimension in the time-space holograms in the universes of the multiverse. Another of Newton's transcripts reads:

NEWTON: And you don't physically live on this simulated world which appears as Earth—you only use it?

SOUL: Yes, that's right, for training purposes.

NEWTON: Why do you call this third sphere the World of Altered Time?

SOUL: Because we can change time sequences to study specific events.

NEWTON: What is the basic purpose of doing this?

SOUL: To improve my decisions for life.[16]

Soul Interventions in Dreams of
Individuals in the Multiverse

Newton's regression/interviews with advanced souls of the "Dreamweaver" category indicate their role in influencing interdimensional consciousness, specifically that of dreams of persons in our universe. Newton writes:

The Dreamweaver souls I have come in contact with all engaged in dream implanting, with two prominent differences.

1. Dream Alteration. Here a skillful discarnate [soul] enters the mind of a sleeper and partially alters an existing dream already in progress. This technique I would call one of inter-lineation, where spirits place themselves as actors between the lines of an unfolding play so the dreamer is not aware of script tampering with the sequences. . . . As difficult as this approach seems, it is evident to me the second procedure is more complex.

2. Dream Origination. In these cases the soul must create and fully implant a new dream from scratch and weave the tapestry of these images into a meaningful presentation to suit their purpose. Creating or altering scenes in the mind of a dreamer is intended to convey a message. I see as this an act of service and love. If the dream implantation is not performed skill-fully to make the dream meaningful, the sleeper moves on and wakes up in the morning remembering only disjointed frag-ments or nothing at all about the dream.[17]

The "Council of Elders" and the Role of Grey Extraterrestrials in a Soul's Reincarnation Cycle

Researcher David Wilcock speculates that the "Council of Elders" that Michael Newton's research found as a consistent feature of the interlife experience may, in fact, be composed of extraterrestrials, perhaps of the Grey exophenotype. Mr. Wilcock writes:

A client of Dr. Newton's reports the Elders as all being hairless, with oval faces, high cheekbones, and smallish features—much like the appearance of certain types of extraterrestrials people have reported seeing. These people did have eyes like ours—not black ovals—

but many extraterrestrial witness reports do feature this type of appearance."[18]

The implication of Mr. Wilcock's speculation, if true, is that the human interlife experience of Newton's subjects would in fact be memories of a deceptive dimensional "virtual experience" in which the presumed Grey extraterrestrials in a dimension of the multiverse posed as advanced spiritual beings in the spiritual dimensions.

On deeper analysis, however, one can reasonably conclude that Mr. Wilcock's speculation may be an unwarranted projection by Wilcock onto the data that Newton has collected about the reported Council of Elders. The full passage by Newton of an individual's soul memory of the interlife that Mr. Wilcock is referring to is as follows:

One of the last requirements before embarkation for many souls is to go before the Council of Elders for the second time. While some of my subjects see the Council only once between lives, most see them right after death and just before rebirth. The spirit world is an environment personified by order and the Elders want to reinforce the significance of a soul's goals for the next life. Sometimes my clients tell me they return to their spirit group after this meeting to say goodbye while others say they leave immediately for reincarnation. The latter procedure was used by a subject who described this exit meeting in the following manner.

"My guide, Magra, escorts me to a soft, white space which is like being in a cloud-filled enclosure. I see my committee of three waiting for me as usual. The middle Elder seems to have the most commanding energy. They all have oval faces, high cheekbones, no hair and smallish features. They seem to me to be sexless—or rather they appear to blend from male to female and back. I feel calm. The atmosphere is formal but not unfriendly. Each in turn asks me questions in a gentle way. The Elders are all-knowing about my entire span of lives but they are not as directive as one might think. They want my input to assess my motivations and the strength of my resolve towards working in new body. I am sure they have had a hand in the body choices I was given for the life to come because I feel they are skilled strategists in life selection. The committee wants me to honor my contract. They stress the benefits of persistence and holding to my values under adversity. I often give in too easily to anger and they remind me of this while reviewing my past actions and reactions towards events and people. The Elders and Magra give me inspiration, hope and encouragement to trust myself more in bad situations and not let things get out of hand. And then, as a final act to bolster my confidence when I am about to leave, they raise their arms and send a power bolt of positive energy into my mind to take with me."[19]

The full range of reported, replicable data about the intelligent civilization of souls that Michael Newton provides is incongruous with the interlife being a deceptive virtual experience staged by Grey or other hyperdimensional extraterrestrials. If anything Newton's data demonstrate that intelligent souls incarnate as Grey extraterrestrial exophenotypes in the various universes of the multiverse.

There is reliable evidence that advanced hyperdimensional extraterrestrials of the Grey exophenotype can undertake an operational role in the reincarnation cycle of intelligent souls from the Grey's platforms in the universes of our multiverse. We will explore this evidence and how one species of Grey hyperdimensionals intervene in a soul's reincarnation cycle in the next chapter.

Parallels between Hyperdimensional Extraterrestrials and Souls in the Time-Space Dimension

Mary Rodwell, who has studied more than 6,000 cases of extraterrestrial contact, has discussed the many parallels that exist between the intervention of hyperdimensional extraterrestrials and the intervention of the intelligent civilizations of souls in the time-space dimension of Earth humans. Ms. Rodwell writes:

> As we explore these potential realities, we find some [Extraterrestrial] encounters suggest that these intelligences may work or interact in dimensions of non-physical realities, such as the dimensions of soul. This encompasses Contact or communications with them, prior to our present physical incarnation, exampled by what Dr. Michael Newton describes in his book, *Destiny of Souls,* the life-between-life, soul state. Furthermore, my research suggests this interaction may also continue after we physically die. Testimony recounts that after our present human incarnation, we may remain involved with these Intelligences . . . Interactions throughout what we would term time and space. I explore some of these extraordinary accounts that are consistent and seem to surface at the deeper level of hypnosis. They illustrate interactions that seem to have no limits in the space-time continuum. I have no way to personally verify any of this data.[20]

8

Hyperdimensional Civilizations' Role in the Omniverse

Role of Hyperdimensional Civilizations in Soul Reincarnation

There is eyewitness evidence that hyperdimensional civilizations based in our universe can play a central role in facilitating soul incarnations into earthly human bodies. The evidence presented here is derived from a case study involving the experiences of Suzanne Hansen, founding director of UFOCUS New Zealand Research Network. She has been a powerful and effective force in bringing about UFO and extraterrestrial disclosure in New Zealand, such as the disclosure of secret files by the Defense Force of New Zealand, which, on Wednesday, December 22, 2010, released more than 2,000 pages of reports by civilians, military personnel, and pilots, detailing unexplained aerial sightings. Throughout her life, Hansen states, her experiences with Greys were "transformational, purposeful, and positive."[1]

Her account is corroborated by Mary Rodwell, RN, who led Hansen through hypnotic regression sessions designed to help her recover her memories of her life-long experiences with Grey hyperdimensionals. Mary Rodwell is a world leader in alien abduction counseling. She was featured in an SBS documentary alongside her son, a self-confessed UFO skeptic. Rodwell, a former nurse, midwife, and clinical hypnotherapist, founded the Australian Close Encounter Resource Network (ACERN) in 1997. She has since investigated more than 1,600 cases of ET encounters worldwide. ACERN helps alien abduction victims deal with their traumas with the support of qualified medical doctors, psychologists, and therapists. In the documentary "My Mum Talks to Aliens," aired nationally in Australia, Rodwell and her son are followed on their journey of discovery as they travel the country in pursuit of solid evidence of an alien presence on Earth. "I believe these advanced intelligences have been visiting this planet since the dawn of time, and I believe they're here to assist evolution to another level," Ms. Rodwell said.[2]

Suzanne Hansen's case reveals an important variation in the dominant method by which souls incarnate in our universe. In it a soul was first introduced to its future Earthling human mother (Hansen) aboard a Grey hyperdimensional spacecraft when she was an eight-year-old child and had been teleported aboard the spacecraft. The mother-to-be and the soul, appearing as a "ball of light," played together under the observation of the Grey hyperdimensionals.

Years later, when the mother was married and pregnant, she was again teleported aboard a Grey hyperdimensional spacecraft and placed on a gurney for a "merging" procedure in which the soul of her future son was "merged" into the fetus then in her womb. The Grey hyperdimensionals told the mother details about the special role her son would play in the future in the time-space dimension on Earth.[3]

All these events took place in the multiverse, not in the interlife dimensions of the Omniverse. This case indicates that under specific circumstances, Grey hyperdimensionals have responsibility for enabling incarnation of souls in the exopolitical dimensions that are normally relegated to guides and the life-selection and preparation process in the interlife.

Hansen's Account of Soul Interactions
on Board Grey Hyperdimensional Craft

In her interview with me, Hansen started by stating that at eight years of age she had been taken by Grey extraterrestrials on board a Grey spacecraft and told that she would be meeting someone who would be "very important" to her in her life. As an adult Hansen began to have flashback experiences of this event and asked Mary Rodwell to do hypnotic regressions to fill in the gaps in her memories. During the regressions with Rodwell, Hansen recovered memories playing on the ship with a group of Grey and transgenic (hybrid) children. All the play was telepathic. She recalled crafting holographic shapes in the air with other children and playing with the shapes as toys. She states that she had had UFO experiences with her family throughout her life, so that this particular experience was not that strange to her.

In preparation for that interview, Hansen sent me a statement, accompanied by three drawings:

(1) During this experience at age eight, I was initially taken to a large room in which I "played" with children of mixed species, using mind games to facilitate telepathic communications between us. The first [drawing] shows the large room . . . I could "see" the objects being produced by our minds, with both my mind (thoughts) and my eyes. In another regression I did with Mary, I described being taught to project a hologram from my mind into the air in front of me. You will notice a strange bed in this drawing, with a baby on it and a Grey beside it. In another regression session, Mary Rodwell regressed me to when I was 6 months old, where I could see myself lying on this small bed. I have nicknamed it the "lullaby bed" and a description of how and why this bed is used with human babies, and the technology involved, features in another chapter of my book.

(2) I was invited to accompany a Grey to a room where I was to

meet another "child." The second drawing shows me as an eight year old, during my first "official" meeting with the ball of blue light, learning to relate to the soul of my future son . . . through play, chasing it around the room while the Greys "observed" from another area of the craft to see if we would be suited.

(3) At age 28, I was taken onboard craft for a "merging" procedure, where the soul of my future son would enter my body, and that of my unborn child, for the first time. The third drawing shows the two entities (mixtures of Grey and another species) comforting me as a Grey begins a calming procedure to lower my metabolism prior to the soul of my son entering my body, and that of the unborn child.

I am very grateful to my husband, who is an artist, for doing draft drawings for my book. I sit with him for hours in order to get them as close as I can recall to the real thing.[4]

In her interview with me Hansen stated that during the visit when she was eight years old, a Grey hyperdimensional who she felt was holding something back approached her and introduced her to a ball of light. She felt confused about the light. The hyperdimensional asked her to look into his eyes. Hansen said that she felt three distinct streams of information coming from him. The hyperdimensional first communicated that the ball of light was a friend, whom she would like to visit. Second, the hyperdimensional communicated a full packet of visual information to her about the life of a child who was the ball of light. Hansen recalls turning cartwheels on the spacecraft room floor as she tried to relate to the ball of light. Finally, the hyperdimensional communicated to her that she and the ball of light had known each other at the soul level for a long time in previous incarnations.*[5]

During the years after this experience, Hansen states she became familiar with the soul state. She became able to communicate with

---

*This description is from an ExopoliticsTV interview I conducted with Hansen, published on January 19, 2011. See notes.

persons who were in the afterlife (or interlife) in the soul state.

It was through her work with Mary Rodwell that Hansen recovered the memory that while pregnant with her son, she was again taken aboard a Grey hyperdimensional craft. On the craft she recalls being escorted by a Grey hyperdimensional (who was familiar because of his "frequency output") down a corridor into a small room where she saw unfamiliar medical equipment. At that point a shorter Grey hyperdimensional, perhaps four feet tall, came into the room, with the archetype of a "short Grey" and different in appearance from the other Greys in the room.

It was at this point that a "merging procedure" began. First, Hansen recalls, the Grey hyperdimensional needed to retrieve some of her earlier memories of interacting with the ball of light. The hyperdimensional placed one hand over her forehead, and Ms. Hansen saw visual memories of her communicating as an eight-year-old child with a blue ball of light on a Grey spacecraft. She became tearful and emotional as she recovered these memories. She was then told that her son's soul had been enhanced as a ball of light, whose signature was more powerful and could harm her if a procedure were not undertaken to lower her metabolism.

Then, Ms. Hansen recalls, about eighteen Grey hyperdimensional children entered the room to observe the soul-merging procedure. She recalls they were chattering away, telepathically. The hyperdimensional told her these children, who were on the way to growing up, were called "Watchers." They appeared to be working with the hyperdimensional on a specific task.

Hansen recalls that at that point her metabolism was lowered. She was barely breathing, and her body temperature was very low. Her voice was slow and quiet. The blue ball of light then entered her body and the body of the fetus of her son in her womb. She recalls that the metabolism of her body was then raised. She was then given information about her son's future and about how the child would be educated both on board the Grey hyperdimensional ship and in the human dimension on Earth. Her son was to receive this dual education for the benefit of mankind, she was told.

Hansen stated that as of 2011, when he was in his mid-twenties, her son had no memory of being aboard the Grey hyperdimensional craft. The Grey hyperdimensionals have told her that when her son is in his mid-thirties, he will recover his memories. She has discussed this history of their prior interaction on the Grey hyperdimensional ship with her son. He now has abilities such as psychic capacity and precognition.

Mary Rodwell's Corroborative Interview

Mary Rodwell's ExopoliticsTV interview with me on January 19, 2011, corroborates Hansen's recollection of her experiences on the Grey ET craft. Rodwell's case materials from her own work as an abduction researcher shed light on how human souls apparently have access to Grey hyperdimensional ships in the form of balls of light.

Rodwell mentions a fifteen-year-old boy who saw two balls of light accompanying him during an encounter with Grey hyperdimensionals. The boy stated that the balls of light were the souls of "Dad and Grandfather," who were at the scene to encourage the boy not to be frightened and to let him know that everything was "OK." Rodwell comments that this case illustrates that Grey hyperdimensionals work with human souls in a very tangible way, although they appear to be intelligences that are different from souls.

Rodwell says her abduction research has led her to conclude that the job of this specific species of Grey hyperdimensionals includes:

1. Gathering genetic information and materials on humans
2. Carrying out healing procedures on humans
3. Cooperation between interdimensional intelligences, as evidenced

by the cooperation between souls and Grey hyperdimensionals

4. Deepening of compassion and love, healing ability, and telepathic transfer of information by souls incarnated as Earthling humans[6]

Of the 1,600 abduction cases Rodwell has investigated, approximately 30 percent involved some level of trauma to the human abductee. The remaining 70 percent of cases do not involve trauma. The medical and health inspections of humans by the Grey hyperdimensionals appear to be for levels of pollution, toxins, and other harmful materials that relate to the planet's problems.

Who Are the Greys and What Is Their Mission?

Hansen states that she knows there are many species of extraterrestrials now visiting Earth. Some of the more than 150 species of Grey hyperdimensionals are working together.

It is a veritable minefield, Hansen says, to understand fully what the mission of the Greys might be, because some people's experiences with Grey hyperdimensionals are fearful, while others are transformational, and so on. Her experiences with the specific species of Grey hyperdimensionals she encountered are purposeful and transformational, although she has had to work through a period of personal fear about the experiences.

Hansen says that her understanding is that souls are universal and can choose to incarnate into a variety of species on different planets. She says that a soul is a spiritual entity, and all of its incarnations appear designed for it to learn moral or evolutionary lessons. She states that she perceives that the Grey hyperdimensionals cannot "command the human race." There is too much chaos on this planet (Earth). The Grey hyperdimensionals' way appears to be to work with souls who are ready to incarnate on a planet and approach its people with love and education. When I asked Hansen if she knows how souls are created and how it is that souls come to be on Grey hyperdimensional craft, she demurs and states she does not know.

Souls Common to
Extraterrestrials and Earthlings

In 2011 I developed a hypothesis like that of researchers Suzanne Hansen and Mary Rodwell, which integrates extraterrestrials and the human soul and is based on empirical data. The hypothesis corroborates and validates their data and findings, as well as the plausibility and veracity of the Hansen case study.[7]

Based on extensive research findings in exopolitics, parapsychology, reincarnation studies, and hypnotic regression, this hypothesis postulates that a civilization of intelligent, evolving souls is apparently created in the spiritual dimensions. As described above, these souls in turn evolve and incarnate into various human or other forms on diverse planets in the known physical universe. In analyzing data from many cases involving hypnotic regression from interlife memories, I extended the hypothesis in order to propose that the community of human souls is reincarnating in complex patterns as extraterrestrials and Earthlings. Hence human souls are common to extraterrestrials and Earthlings.

Both the Suzanne Hansen case and the cases reported on by Michael Newton support this hypothesis. The Hansen case demonstrates that souls and specific Grey hyperdimensionals navigate the multiverse as interdimensional entities in cooperative relationships, such as mutual assistance in education of a future parent of an incarnating soul and in the actual incarnation process. Thus some souls, such as those of Suzanne Hansen's future son and those of the reported father and grandfather of the fifteen-year-old boy who accompanied him during an encounter with Grey hyperdimensionals, appear to be part of a structured program of cooperation among the intelligent civilizations of souls and this specific species of Grey hyperdimensionals. The program includes specialized intervention in specific types of soul incarnations and educational encounters with Grey hyperdimensionals in the exopolitical dimension.

At least one organization researching the role of Grey hyperdimensional extraterrestrials has preliminarily concluded that a specific species of Grey extraterrestrials may be carrying out an extensive intervention in the Earth human reincarnation cycle.[8] Earth humans are at the beginnings of our collective understanding of the full dimensional factors that may be affecting the human soul's reincarnation cycle. This is a priority area for our further research, understanding, and species' cosmic growth.

The Science of the Omniverse

At least ten implications for further exploration of the dimensional ecology of the Omniverse hypothesis and for a positive human future can be drawn from this research into the science of the Omniverse.

1. Evidence Supports the Dimensional Ecology of the Omniverse Hypothesis

A reasonable observer can now conclude that presumptive empirical evidence supports the dimensional ecology of the Omniverse hypothesis, which holds that we Earthlings live in an Omniverse that encompasses intelligent civilizations based in both the exopolitical dimensions or the multiverse and in the spiritual dimensions. The totality of the spiritual dimensions (souls, spiritual beings, and God itself) functions as the Source of the universes of the exopolitical dimensions.

2. Exopolitics and Parapsychology

Exopolitics, the science of relations among intelligent civilizations, and parapsychology, the science of psi consciousness, telepathy, reincarnation, the soul, the interlife, and Source (God), are among the proper scientific disciplines for exploring and mapping the dimensional ecology of the Omniverse.

It is time that exopolitics and parapsychology be admitted to the accepted canon of sciences and openly taught at all levels of education: elementary, secondary, and postsecondary. The state censorship barriers of official ridicule and denial that were constructed in 1953 by the U.S. Central Intelligence Agency in its Robertson Panel Durant Report against any civil society discourse, publication, research, or public knowledge of a populated, organized Omniverse must now be forever deconstructed and discarded. The governments and intelligence communities, along with the financial and political elites that control them, do not have the spiritual, moral, or legal authority to suppress research and access to the Omniverse, including the exopolitical dimensions, the interlife, the spiritual dimensions, and Source (God).[1]

3. Dimensionality Is a Key Design Criterion of the Omniverse

Dimensionality, the ability of intelligence to organize itself via dimensions (discrete bands of conscious energy), appears to be a key criterion by which the Omniverse is designed, both in the spiritual dimensions and in the exopolitical dimensions. Intelligent civilizations (including extraterrestrial and interdimensional ones), in both the spiritual and the exopolitical dimensions, can be typed by the dimension in which they are based.

4. Human Soul Consciousness and Development

The consciousness and developmental level of souls incarnating on Earth and our collective ability to comprehend the dimensional ecology of the multiverse is both a bottleneck and a key to the future evolution of Earthling humans in the society of organized intelligent life in the exopolitical dimensions.

Commenting on his findings about how souls are distributed by level of development (beginner, intermediate, advanced souls), Michael Newton writes, "I believe almost three-quarters of all souls who inhabit human bodies on Earth today are still in the early stages of development. I know this is a grossly discouraging statement because it means most of the human population is operating at the lower end of their training. . . . Nevertheless, I see the possibility that we may have only a few hundred thousand people [souls] on Earth at level V [Highly Advanced]."[2]

5. Positive Transformation of Power, Economic, and Social Structures on Earth

The long-term positive transformation of power, economic, and social structures on Earth may depend on "soul power." Significant soul development during an incarnation can occur as science-based knowledge about the dimensional ecology of the Omniverse is seriously internalized by the more advanced and intermediate souls incarnating on Earth, and presented to the approximately three-quarters of the Earth's population who incarnate beginner's souls (assuming that Newton's findings are empirically correct), in a context in which hyperdimensional civilizations support an expansion of Earthling human consciousness.

Carl Johan Calleman, an author and researcher who has developed the unity consciousness theory of the Mayan Calendar, has hypothesized that from October 28, 2011, onward, the singularity or interdimensional portal at the center of our galaxy (the galactic center "black hole") has been emanating constant "enlightened universal Unity consciousness." According to Calleman the singularity or interdimensional portal modulates this universal unity consciousness energy wave as well. This is, by any standard, exceedingly good scientific news.

Universal, alternating energy-wave movements have been a feature of our universe, says Calleman, since the Big Bang, and it is these wave movements that have shaped the nature of consciousness in our universe over the past 18 billion years.[3]

In my interview with him, Calleman stated, "The universal, alternating energy-wave movements [have ended], and Earth is set on a gradual setting of a potential to reach advanced utopian planet status, a virtual 'Garden of Eden.'"[4] This milestone marks the beginning of final preparations for the entry of Earth into a long-prophesized Golden Age of Utopia, or "positive timeline."

What, you may ask, does this universal wave of unity consciousness have to do with you or your reality? In a word, everything.

The wave of unity consciousness, like all universal energy-wave emanations, creates the "meme" or story content of our personal and collective reality. Calleman's discoveries suggest that our highest thoughts and "memes" may in fact be sourced from the energy waves of the universal singularity, mediated in our universe via the galactic singularity, the black hole at the center that the Mayans called Hunab Ku, and in our solar system via its singularity—Sol, our sun—which is a dimensional portal to other galaxies, according to physicist Nassim Haramein.[5]

These universal singularity energy waves serve as carrier waves for "the universe's mind and spirit software." That is, they are the way that an intentional universe (as created and maintained by a collective, conscious Source) lets us know what it intends for consciousness in our entire universe. As a practical matter does it matter if you are tuned in to and committed to the universal energy wave of unity consciousness? Yes, the universal energy waves we humans tune in to and commit to determine our planetary status and how we reach our potential as a planet.

Will Earth be a dystopia or a utopia? The good news is that our galactic singularity is now emanating "utopia" consciousness waves on its alternating energy carrier waves. The more we humans can individually and collectively tune in to the universal energy wave of "enlightened universal consciousness," and can commit to achieving nondualistic consciousness, then the stronger the "positive timeline" becomes, and the more rapidly we will achieve our potential as a planet to reach advanced utopian planet status.

Extraterrestrial researcher Mary Rodwell shared details of a coming global shift that she has discovered in her investigations of conscious interaction with extraterrestrials, including at the soul level. The coming shift is described as including an awakening and transformation of human consciousness, along with dramatic changes in the way human society functions on the Earth.[6] Some researchers state that this global shift of consciousness is a long-term dimensional project conducted by Grey hyperdimensionals and may be limited to the contactees and abductees of the Greys.

6. Comprehending Extraterrestrial, Hyperdimensional, and Exopolitical-Related Phenomena and Information

The dimensional ecology of the Omniverse hypothesis and the dimension-based typology of intelligent civilizations in the exopolitical dimensions can facilitate and accelerate the comprehension of UFO and extraterrestrial-related data and information by the Earthling human public and by scientific, governmental, educational, and media organizations. It can thus accelerate the pace of extraterrestrial disclosure, including disclosure by these intelligent civilizations themselves.

7. Comprehension of the Spiritual Dimension

Nature of the Soul, Interlife, God

The dimensional ecology of the Omniverse hypothesis and the science-based study of the spiritual dimension can facilitate and accelerate humanity's comprehension of true versions of the basic concepts of reality, such as soul, interlife, and Source (God), by the Earthling human public and by scientific, governmental, educational, and media organizations. Open questions, such as whether more than one Omniverse and one Source (God) exist, can be addressed. This will accelerate the pace of conscious evolution of Earthling humanity and of the population of souls incarnating on Earth.

8. The Centrality of the Intelligent Civilizations of Souls to the Universes of the Exopolitical Dimensions

The soul—which is a holographic fragment of the Creator Source (God)—is a common denominator to all the dimensions of the Omniverse. The science-based study of the exopolitical and spiritual dimensions of the Omniverse reveals the centrality of the intelligent civilizations of souls in creation, maintenance, design, and ultimately incarnation in the universes of the exopolitical dimensions.

The human soul inhabits the spiritual dimensions between physical lives and inhabits the exopolitical dimensions during lives (and may lead several exopolitical lives simultaneously). Exopolitical lives may be as a human on Earth or an intelligent being on another planet or in another universe or dimension. There may be a multiplicity of types of intelligent souls that our science will discover.

There is preliminary contactee witness evidence that a specific species of Grey hyperdimensionals may have an extensive role in the reincarnation cycle of Earth humanity's souls. This is a vital area for future research and for humanity's understanding of its incarnating on Earth.

9. Education about the Soul, Interlife, and Source (God)

Humanity now is being misinformed about the true nature of the soul, the interlife, the mechanisms of reincarnation, and ultimately of

Source (God). Religions are a large source of erroneous information about these realities. This misinformation is based on texts and ancient religious belief systems that are not scientifically correct and are yet considered sacred, as a matter of faith.

Earth religions have resulted from manipulatory extraterrestrial intervention, such as the Anunnaki extraterrestrial intervention on Earth starting approximately 280,000 years ago.[7] Academic science prohibits teaching the reality of life after death, even though empirical evidence supports this hypothesis. Therefore, modern science is also a source of erroneous information about the soul, the interlife, and Source (God). Cultural and religious conflicts result from the ignorance imposed by religion and an antiquated scientific canon concerning any such spiritual realities. The science of the spirit can demonstrate the true nature of the human soul and its spiritual and exopolitical dimensions.

10. Science and Spirituality

The dimensional ecology of the Omniverse hypothesis is itself a matrix for recognition and classification of ongoing and new research into intelligent civilizations in the exopolitical and spiritual dimensions. This hypothesis represents a coming together of science and spirituality that will return science to the proper study and understanding of the human soul, and a return to the supremacy of understanding over ignorance. To support and encourage this understanding is our duty as informed, aware human souls.

Notes

Preface.
The Beginning of This Book

1. Webre, *Exopolitica: La Politica, El Gobierno y La Ley en el Universo.*

Introduction. A Map of the Omniverse

1. Markowitz, "Five Reasons We May Live in a Multiverse," http://bit
.ly/18Q7t8N (accessed May 3, 2015).
2. Amoroso and Rauscher, *The Holographic Anthropic Multiverse,* 32.
3. Nancy Atkinson, "If We Live in a Multiverse, How Many Are There?"
www.universetoday.com/42696/if-we-live-in-a-multiverse-how-many-are-
there/ (accessed June 25, 2015); Andrei Linde and Vitaly Vanchurin, "How
Many Universes Are in the Multiverse?" http://arxiv.org/pdf/0910.1589v3
.pdf (accessed May 3, 2015).
4. Ibid.
5. Krauss, *A Universe from Nothing,* www.pdf-archive.com/2014/10/19/
lawrence-m-krauss-a-universe-from-nothing/lawrence-m-krauss-a-universe-
from-nothing.pdf (accessed June 30, 2015), 19.
6. "Ecology," Wikipedia, https://en.wikipedia.org/wiki/Ecology (accessed
June 25, 2015).
7. Helmenstine, "Scientific Hypothesis, Theory, Law Definitions," http://bit
.ly/bV90Rj (accessed May 3, 2015).
8. "Evidence," Wikipedia, http://bit.ly/1ETodZS (accessed June 30, 2015).
9. Newton, *Destiny of Souls,* 127–28.

Chapter 1.
Understanding the Omniverse

1. Kramer, "Scale of Universe Measured," http://bit.ly/1e96Rhq (accessed May 3, 2015).

2. "500 Billion—A Universe of Galaxies," http://bit.ly/1grvKpP (accessed May 3, 2015).

3. Sebastian Anthony, "Astronomers Estimate 100 Billion," http://bit.ly/13UiMgU (accessed May 3, 2015).

4. Randy Krum, "How Many Alien Civilizations?" http://bit.ly/16ygWlH (accessed May 3, 2015).

5. Report of Scientific Advisory Panel, http://bit.ly/15ry5Na (accessed May 3, 2015).

6. Troy Matthew, "Science or Sacrilege? Atheists and Agnostics Are 76% More Likely Than Christians to Believe in the Existence of Extraterrestrial Life," http://bit.ly/17omJWM (accessed May 3, 2015).

7. "One in Five (20%) Global Citizens Believe That Alien Beings Have Come Down to Earth and Walk Amongst Us in Our Communities Disguised as Humans," Ipsos.com, http://bit.ly/1cx0qWP (accessed May 3, 2015).

8. Newton, *Life Between Lives, Journey of Souls, Destiny of Souls.*

9. "Ipsos Global Advisory: Supreme Being(s), the Afterlife and Evolution," Ipsos.com, http://bit.ly/19mrjcT (accessed May 3, 2015).

10. "Americans' Belief in God, Miracles and Heaven Declines," Harris Interactive Polls, harrisinteractive.com, http://bit.ly/1dne4bJ (accessed May 3, 2015).

11. Ibid.

12. Ibid.

13. Ibid.

14. Ibid.

15. Clow and Clow, *Alchemy of Nine Dimensions,* xxiii.

16. Goswami, *The Self-Aware Universe,* 9–10.

17. Bromberg, "On Goswami's Monistic Idealism Worldview," 11.

18. James S. Gordon, "The UFO Experience," http://bit.ly/1fM1yHn (accessed May 3, 2015).

19. Alfred Webre, "Exopolitics: A Decade of Contact—Part 2," UFO Evidence .org, http://bit.ly/GVR1dH (accessed May 3, 2015).

20. "Associated Press Reports that 'EXOPOLITICS' Nominated for 2005

'Word of the Year,'" Exopolitics: Politics, Government, and Law in the Universe, http://bit.ly/HaArqd (accessed May 3, 2015).

21. Webre, *Exopolitics,* 92.

22. "Project Pegasus," www.facebook.com/groups/ProjectPegasus/search/?query =montauk%20chair# (accessed June 30, 2015).

23. Webre, "Third Whistle Blower Confirms Obama's Participation in CIA Jump Room Program of Early 1980's," http://bit.ly/O5TRcO (accessed June 30, 2015).

24. Ron Cowen, "Simulations Back up Theory That Universe Is a Hologram," Nature online, http://bit.ly/1bDPwvE (accessed May 3, 2015).

25. Talbot, *The Holographic Universe;* Barrett, *The Holographic Canvas.*

26. Kate Becker, "Searching for Advanced Alien Engineering," The Nature of Reality, www.pbs.org/wgbh/nova/blogs/physics/2015/05/searching-for-advanced-alien-engineering/ (accessed October 6, 2015).

27. Danielle Wiener-Bronner, "Don't Freak Out, but the Universe May Be a Hologram," http://fus.in/1JV4nBJ (accessed June 30, 2015).

28. Webre, "Web Bot: Andrew Basiago Is Predicted 'Planetary Level' Whistleblower for Mars Life and Time Travel," http://exopolitics.blogs .com/exopolitics/2011/12/web-bot-andrew-basiago-is-predicted-planetary-level-whistleblower-for-mars-life-and-time-travel.html (accessed June 25, 2015); Webre, "Why Is Hawking Affirming Time Travel and Appearing Ignorant of DARPA Secret Time Travel?" examiner.com, http://exm.nr/ GO2Cvz (accessed May 3, 2015); Webre, "Second Whistleblower Emerges to Confirm Reality of Time Travel," www.bibliotecapleyades.net/ciencia/ time_travel/esp_ciencia_timetravel33.htm (accessed June 25, 2015).

29. Newton, *Destiny of Souls,* 127–28.

30. Webre, "Time Travel and Political Control," Exopolitics: Politics, Government, and Law in the Universe, http://bit.ly/LoBF0T (accessed May 3, 2015).

31. Webre, "My 1970s Meeting with DARPA's Project Pegasus Secret Time Travel Program," Exopolitics: Politics, Government, and Law in the Universe, http://bit.ly/182i3dE (accessed May 5, 2015).

32. "The Urantia Book," Urantia Foundation, http://bit.ly/1fCFBIr (accessed May 3, 2015).

33. Wyllie, *Revolt of the Rebel Angels,* 4.

34. Andrei Linde and Vitaly Vanchurin, "How Many Universes Are in the

Multiverse?" Cornell University Library online, http://bit.ly/HtbdEr (accessed May 3, 2015).

Chapter 2.
Typology of Intelligent Civilizations

1. Personal correspondence from Oxford University Press to author, August 2009.
2. "Typology," Wikipedia, http://bit.ly/1bbmDDI (accessed May 3, 2015).
3. Alfred Webre, "Exopolitics Researcher Develops Evidence-based Typology of Extraterrestrial Civilizations," examiner.com, http://exm.nr/1fy4bwn (accessed May 3, 2015).
4. Joe Garland, "The Kardashev Scale," Quarks to Quasars, http://bit. ly/1bNRJBm (accessed May 3, 2015); George Dvorsky, "How to Measure the Power of Alien Civilizations Using the Kardashev Scale," i09, http:// bit.ly/J3G0Yy (accessed May 3, 2015).
5. Michael Salla, "A Report on the Motivations and Activities of Extraterrestrial Races," Exopolitics.org, http://bit.ly/H0ajyp (accessed May 3, 2015).
6. Wilcock, *The Synchronicity Key*, 9–10.
7. Elkins et al., "The RA Materials," verified at The Law of One, http://www. lawofone.info/; The Law of One, www.llresearch.org/library/the_law_of_ one_pdf/the_law_of_one_pdf.aspx (accessed June 25, 2015).
8. Ibid.
9. Ibid.
10. Ibid.
11. Ibid.
12. Ibid.
13. Ibid.
14. Ibid.
15. Manuel Lamiroy, "Density," Exopaedia, http://bit.ly/1cF2Z9f (accessed May 9, 2015).
16. Newton, *Destiny of Souls*, 127–8.
17. Pereira, *Arcturan Star Chronicles*.
18. Brown, *Cosmic Explorers*, 147–77.
19. Lamiroy, "Exophenotypes," Exopolitics South Africa, http://bit.ly/1giGxFn (accessed May 3, 2015).

20. Ibid.

21. Ibid.

22. Lamiroy, "Non-humanoid Mammalian Phenotype," Exopaedia, http://bit. ly/16RSNXn (accessed May 3, 2015).

23. Lamiroy, "Humanoid Phenotype," Exopaedia, http://bit.ly/19PpukK (accessed May 3, 2015).

24. Andrew D. Basiago, Exopolitics Mars, http://on.fb.me/1kjHfUD (accessed June 25, 2015).

25. Lamiroy, "Grey Phenotype," Exopaedia, http://bit.ly/19PpukK and http:// bit.ly/H6Dksh (accessed May 3, 2015).

26. Ibid.

27. "The Exobiology Project: Talking with the Small Greys," Open Minds Forum 2013, http://openmindsforum.forumotion.com/t160-the-exobiology-project-talking-with-small-greys (accessed May 13, 2015).

28. Lamiroy, "Reptilian Phenotype," Exopaedia, http://bit.ly/1czbS1F (accessed May 13, 2015).

29. "The Exobiology Project: Talking with the Hooded Reptilian," Open Minds Forum 2013, http://openmindsforum.forumotion.com/t160-the-exobiology-project-talking-with-small-greys (accessed May 13, 2015).

30. Lamiroy, "Insectoid Phenotype," Exopaedia, http://bit.ly/19ZhN1u (accessed May 3, 2015).

31. Lamiroy, "Hybrid Phenotype," Exopaedia, http://bit.ly/17VJEc5 (accessed May 3, 2015).

32. Huyghe, *Field Guide to Extraterrestrials;* Dickinson and Schaller, *Extraterrestrials: A Field Guide for Earthlings.*

33. "General/Mass Sightings," UFO Evidence, http://bit.ly/18RXnFH (accessed May 13, 2015).

34. "Humanoid Cases," UFO Evidence, http://bit.ly/19Ogyxj (accessed May 13, 2015).

35. UFO Evidence, www.ufoevidence.org/topics/PilotSightings.htm (accessed June 25, 2015).

36. Lamiroy, "Exophenotypes," http://bit.ly/1giGxFn (accessed May 13, 2015).

37. Webre, "Report: Between 2% of US Population and 14.7% of Global Population Are Being Involuntarily Teleported (Abducted) by Hyperdimentional Civilizations," http://exopolitics.blogs.com/exopolitics /2013/08/report-between-2-of-us-population-and-147-of-global-population-

are-being-involuntarily-teleported-ab.html (accessed June 25, 2015).

38. Ibid.

39. Webre, "Mary Rodwell—ETs, Souls, The New Humans and a Coming Global Shift," http://bit.ly/17tBY7B (accessed May 13, 2015).

Chapter 3.
Time-Space Solar-System Civilizations

1. "What Does It Mean to be Human?" Smithsonian National Museum of Natural History, http://bit.ly/GYzJfW (accessed May 13, 2015).

2. Alfred Webre, "New Data, Law of Evidence Support View of Mars Having Indigenous, Intelligent Extraterrestrial Life," Exopolitics: Politics, Government, and Law in the Universe, http://bit.ly/P9nxw6 (accessed May 13, 2015).

3. Andrew Basiago, "Bernard Mendez," Project Pegasus, http://on.fb.me/1adQyyt (accessed May 13, 2015).

4. Matt Slick, "Extraordinary Claims Require Extraordinary Evidence," CARM, http://bit.ly/1aAH796 (accessed May 13, 2015).

5. Basiago, "Project Pegasus," Project Pagasus, http://on.fb.me/19pjqmo (accessed May 13, 2015).

6. Basiago, "Five Exophenotypes of Bipedal Humanoids," Project Pegasus, http://on.fb.me/1kjHfUD (accessed May 13, 2015); "Andrew D. Basiago," Mars Anomaly Research Society, http://bit.ly/HbmVDJ (accessed May 13, 2015).

7. Personal communication, conversation with Andrew D. Basiago regarding his telephone conversation with Virginia Olds.

8. Basiago, "Five Exophenotypes of Bipedal Humanoids."

9. Webre, "New Data, Law of Evidence Support View of Mars Having Indigenous, Intelligent Extraterrestrial Life," Exopolitics: Politics, Government, and Law in the Universe, http://bit.ly/P9nxw6 (accessed May 13, 2015).

10. Basiago, "Discovery," http://bit.ly/1aALaSK (accessed June 25, 2015).

11. Basiago, "Three Principal Exophenotypes of Humanoid Beings on Mars by Andrew D. Basiago," http://bit.ly/1cahkeq (accessed May 13, 2015).

12. Mike Spent, "Mars: Andrew D. Basiago," http://bit.ly/1aGBdTK (accessed May 3, 2015).

13. Ibid.

14. Webre, "New Data, Law of Evidence."

15. Ibid.

16. Webre, "Mars Visitors Basiago and Stillings Confirm Barack Obama Traveled to Mars," Exopolitics: Politics, Government, and Law in the Universe, http://bit.ly/10L2P5t (accessed May 13, 2015); Webre, "Third Whistleblower Confirms Obama's Participation in CIA Jump Room Program of Early 1980s," Exopolitics, http://bit.ly/O5TRcO (accessed May 13, 2015).

17. Webre, "Mars Visitors."

18. Ibid.

19. Ibid.

20. Ibid.

21. Basiago, "Mars," Fear & Loathing online, http://bit.ly/1aGBdTK (accessed June 25, 2015), 43.

22. Webre, "Mars Visitors."

23. Ibid.

24. Michael Rundle, "White House Denies President Obama Travelled to Mars via Teleport at Age 19," Huffington Post, http://huff.to/1fU5xla (accessed May 13, 2015).

25. Webre, "Third Whistleblower."

26. Basiago, "Bernard Mendez," Project Pegasus, http://on.fb.me/1adQyyt (accessed May 13, 2015).

27. Webre, "Third Whistleblower."

28. Ibid.

29. Ibid.

30. Ibid.

31. Ibid.

32. Ibid.

33. Ibid.

34. Ibid.

35. Webre, "Whistleblower Laura Magdalene Eisenhower, Ike's Great-Granddaughter, Outs Secret Mars Colony Project," http://exopolitics.blogs.com/exopolitics/2011/12/by-alfred-lambremont-webre-jd-med-laura-magdalene-eisenhower-ikes-great-granddaughter-in-a-public-statement-lau.html (accessed June 30, 2015).

36. Michael Salla, "Whistleblower Claims He Served 17 Years at Secret Mars Military Base," Exopolitics.org, http://bit.ly/1puQO4D (accessed May 13, 2015).

37. "Mars Experiencer Andrew D. Basiago Validates Mars Experiences of U.S. Marine Capt. Randy Cramer," Exopolitics, http://bit.ly/1pv0exc (accessed May 13, 2015).

38. Ibid.

39. "Mars Exploration Program," NASA Mars Exploration, http://1.usa .gov/16wjz3Z, (accessed May 13, 2015).

40. Basiago, "The Discovery of Life on Mars," Mars Anomaly Research Society, http://bit.ly/1aALaSK (accessed May 13, 2015); Mars Anomaly Research Society (MARS), www.projectmars.net; "Martian Humanoid Exophenotypes," exopolitics.blogs.com, http://bit.ly/19Noxgb (accessed May 14, 2015).

41. Basiago, "The Discovery of Life on Mars," http://bit.ly/1aALaSK (accessed May 13, 2015), 6.

42. Ibid., 1.

43. "Martian Humanoid Exophenotypes," exopolitics.blogs.com, http://bit .ly/19Noxgb (accessed May 14, 2015).

44. "PIA10214, Spirit's West Valley Panorama," NASA, Jet Propulsion Laboratory Photojournal, http://1.usa.gov/18Ue2Iu (accessed May 14, 2015).

45. Basiago, "Hoagland Affirms Belief That No Evidence Exists Mars Is Inhabited," http://bit.ly/16yOv9f, Mars Anomaly Research (accessed May 14, 2015).

Chapter 4. Hyperdimensional Civilizations

1. Alfred Webre, "Up to 1 Billion Humans," www.bibliotecapleyades. net/vida_alien/alien_abductionabductees30.htm; Webre, "Intention Experiments, Interdimensional UFOs, and Multi-dimensional Beings Converge at Mt. Adams, WA," examiner.com, http://exm.nr/18DT1gZ (accessed May 14, 2015).

2. Webre, "Report: Between 2% of US Population and 14.7% of Global Population Are Being Involuntarily Teleported (Abducted) by Hyperdimentional Civilizations," Exopolitics: Politics, Government, and Law in the Universe, http://bit.ly/1f4GcCi (accessed May 9, 2015).

3. "Alien Abduction Claimants," Wikipedia, The Roper Poll, http://bit .ly/195jNAh (accessed May 14, 2015).

4. Webre, "Mary Rodwell—ETs, Souls, The New Humans and a Coming Global Shift," Exopolitics: Politics, Government, and Law in the Universe, http://bit.ly/17tBY7B (accessed May 14, 2015).

5. Neil Gould, "Galactic Diplomacy: The Greys," ExoNews, http://bit. ly/181e3qD (accessed May 14, 2015); Neil Freer, "Sapiens Rising," http:// bit.ly/19H6gTj (accessed May 15, 2015).

6. International Community for Alien Research (ICAR), http://bit .ly/1aEN6v7 (accessed May 14, 2015).

7. Webre, "Report: Between 2%."

8. "Alien Abduction," Wikipedia, http://bit.ly/HqHEmX (accessed May 15, 2015).

9. Webre, "Report: Between 2%."

10. Webre, "When Will Extraterrestrial 'Disclosure,' or Contact in the Public Domain Happen?" Exopolitics: Politics, Government, and Law in the Universe, http://bit.ly/1gdjmLC (accessed May 15, 2015).

11. "War," Wikipedia, http://bit.ly/16mWQKW (accessed May 14, 2015).

12. Basiago, "Project Pegasus," ProjectPegasus.net, http://bit.ly/1aUmkxp (accessed May 14, 2015).

13. Webre, "Basiago and Eisenhower Reveal 'Marsgate' and Make Case for 'Alternative 4,'" examiner.com, http://exm.nr/17ya8VD (accessed May 15, 2015).

14. Webre, "Time Travel and Political Control," Exopolitics: Politics, Government, and Law in the Universe, http://bit.ly/LoBF0T (accessed May 14, 2015).

15. Relfe, The Mars Records, Vol. 2, 204.

16. Ibid., 205.

17. Ibid., 203.

18. Ibid., vol. 1, 65, 77.

19. Ibid., vol. 2, 110.

20. Webre, "Archons—Exorcising Hidden Controllers with Robert Stanley and Laura Eisenhower," http://exopolitics.blogs.com/exopolitics/2011/12/ archons-exorcising-hidden-controllers-with-robert-stanley-and-laura- eisenhower.html (accessed May 14, 2015).

21. Ibid.

22. Ibid.

23. John Lash, "Who Wrote the Reptilian Agenda?" UFO Digest, http://bit .ly/1cLWRXT (accessed May 14, 2015).

24. Robinson et al., *NHC II*, 1:28.16 as cited in Webre, "Archons—Exorcising Hidden Controllers with Robert Stanley and Laura Eisenhower," http:// exopolitics.blogs.com/exopolitics/2011/12/archons-exorcising-hidden-controllers-with-robert-stanley-and-laura-eisenhower.html (accessed June 26, 2015).

25. See Australian Close Encounter Research Network (ACERN), http://bit .ly/18DVPed (accessed May 14, 2015).

26. Mary Rodwell, "Extraterrestrials, Human Consciousness and Dimensions of Soul: The Intimate Connection," Exopolitics: Politics, Government, and Law in the Universe, http://bit.ly/19TrVWY (accessed May 14, 2015).

27. Ibid.

28. Ibid.

29. Ibid.

30. Ibid.

31. Ibid.

Chapter 5. Intelligent Civilizations' Governance Authorities

1. Andrei Linde and Vitaly Vanchurin, "How Many Universes Are in the Multiverse?" Cornell University Library online, http://bit.ly/HtbdEr (accessed May 14, 2015); Randy Krum, "How Many Alien Civilizations Are There in the Galaxy?" Cool Infographics.com, http://bit.ly/16ygWlH (accessed May 14, 2015).

2. Lindy Newlove-Eriksson and Johan Eriksson, "Governance beyond the Global: Who Controls the Extraterrestrial?" Allacademic Research online, http://bit.ly/1g6woaX (accessed May 14, 2015); C. S. Cockell, "Liberty and the Limits to the Extraterrestrial State," JBIS online, http://bit.ly/1crfovd (accessed May 14, 2015).

3. Arjun Walla, "Russian Prime Minister Confirms the Existence of Intelligent Extraterrestrial Life," Collective Evolution, http://bit.ly/1hYqkT4 (accessed May 15, 2015).

4. Brown, *Cosmic Explorers,* 147–77.

5. "What Is Remote Viewing?" The Farsight Institute, http://bit.ly/1iB4cNb (accessed May 15, 2015).

6. Brown, *Cosmic Voyage,* 92.

7. Webre, "Author: ET Council Seeded *Homo sapiens* as Intelligent Beings with 12-strand DNA," examiner.com, http://exm.nr/1h6KiOF (accessed May 15, 2015).

8. Webre, "ET Council: We Will Increase UFOs, Address U.N. in 2014, Renew Ecology in 2015," examiner.com, http://exm.nr/HtiNyQ (accessed May 15, 2015); "More Predicted UFO Sightings over New York Confirm ET Will Intervene in Ecology," examiner.com, http://exm.nr/17CmVq5 (accessed May 15, 2015); "Stanley Fulham Dies, Warned ETs," examiner. com, http://exm.nr/16T0j6m (accessed May 15, 2015).

9. Webre, "ET Council."

10. Webre, "More Predicted UFO Sightings over New York Confirm ET Will Intervene in Ecology," http://ufodigest.com/article/more-predicted-ufo-sightings-over-new-york-confirm-ets-will-intervene-earths-ecology (accessed June 26, 2015).

11. Webre, "ET Council."

12. Webre, "Are the UFOs Appearing in Russia Skies in Dec. 2010 Fulham's Predicted UFO Wave?" examiner.com, http://exm.nr/17y2DQW.

13. Ibid.

14. Fulham, *Challenges of Change.*

15. Personal communication.

16. Webre, "January 2011 UFO Wave over Moscow Is 3rd Independent Confirmation of ET Council," www.bibliotecapleyades.net/disclosure/disclosure34.htm (accessed June 26, 2015).

17. Webre, "More Predicted UFO Sightings"; Webre, "January 2011 UFO Wave."

18. Personal communication, e-mail received from Expert Randy Kitchur.

19. "Jewish Rabbi Details Possible Hebrew Calendar Rapture Timeline," http://bit.ly/1ctlzyU (accessed May 15, 2015).

20. Webre, "Part 2—Peter Kling: Christ Returns As an Extraterrestrial with the Armies of the Multiverse to Defeat the NWO?" Exopolitics, http://bit.ly/JSviEI (accessed May 15, 2015).

21. Webre, "UFO Expert: Stanley Fulham UFO Predictions Fulfilled over NY, Moscow, and London," www.examiner.com/article/ufo-expert-stanley-

fulham-ufo-predictions-fulfilled-over-ny-moscow-and-london#ixzz1E0bhe8ir (accessed June 30, 2015).

22. Ibid.

Chapter 6. The Interlife Dimensions

1. Senkowski, "Instrumental Transcommunication ITC," http://itcvoices .org/wp-content/uploads/2015/04/Instrumental-Transcommunication-Senkowski-Full.pdf (accessed June 26, 2015).

2. Ibid.

3. Zammit and Zammit, *A Lawyer Presents the Evidence,* 165.

4. Ervin Lazlo, "An Unexplored Domain of Nonlocality," www.worlditc .org/f_21_laszlo_scientific_explanation_of_itc.htm (accessed June 26, 2015).

5. Senkowski, "Instrumental Transcommunication ITC," http://itcvoices .org/wp-content/uploads/2015/04/Instrumental-Transcommunication-Senkowski-Full.pdf (accessed June 26, 2015).

6. Zammit and Zammit, *A Lawyer Presents the Evidence,* 158.

7. Ibid., 158; Kubris and Macy, *Conversations Beyond the Light,* 102.

8. Ibid., 159.

9. Ibid., 159; Ostrander and Schroeder, *Psychic Discoveries Behind the Iron Curtain.*

10. Ostrander and Schroeder, *Psychic Discoveries Behind the Iron Curtain*; Zammit and Zammit, *A Lawyer Presents the Evidence,* 158–59.

11. Basiago, "The Sources of Time Travel," http://bit.ly/1cPmciU (accessed May 9, 2015).

12. Mark Williams, "Invention of a Machine for Photographing the Past," Project Pegasus, http://on.fb.me/HykL0u (accessed May 15, 2015).

13. Kevin Williams, "Dr. Ian Stevenson's Reincarnation Research," http://bit .ly/HKqjF9 (accessed May 15, 2015).

14. Ibid.

15. Ibid.

16. Fontana, *Is There an Afterlife?*

17. Ring, *Lessons from the Light;* Moody, *Life after Life.*

18. Zammit and Zammit, *A Lawyer Presents the Evidence*, 22–30.

19. Ibid., 111.

20. Berman and Lanza, *Biocentrism,* 159.

21. Anna Lemind, "Quantum Theory Proves That Consciousness Moves to Another Universe after Death," Learning Mind, http://bit.ly/1f6wQFk (accessed May 15, 2015).

Chapter 7.
The Intelligent Civilizations of Souls

1. Peake, *Is There Life After Death?* 5039–40.

2. Tim Miejan, "Journey of Souls with Michael Newton," the Edge, http://bit.ly/1bpsY2m (accessed May 15, 2015).

3. Newton, *Destiny of Souls,* 1566–67.

4. Newton, *Journey of Souls,* 127.

5. Newton, *Destiny of Souls,* 2261–69.

6. Pereira, *Arcturan Star Chronicles.*

7. Newton, *Journey of Souls,* 103.

8. Newton, *Destiny of Souls,* 86.

9. Ibid., 340.

10. Newton, *Journey of Souls,* 261–65, 287–94, 294–96.

11. Ibid. 2325–26.

12. Ibid.

13. Ibid., 2330–34, 2339–41.

14. Newton, *Destiny of Souls,* 5615–20.

15. Ibid., 354.

16. Newton, *Journey of Souls,* 160.

17. Newton, *Destiny of Souls,* 619–34.

18. Wilcock, *The Synchronicity Key,* 194–95.

19. Newton, *Journey of Souls,* 3170–76.

20. Mary Rodwell, "Extraterrestrials, Human Consciousness and Dimensions of Soul: The Intimate Connection," http://bit.ly/19TrVWY (accessed May 15, 2015).

Chapter 8. Hyperdimensional
Civilizations' Role in the Omniverse

1. Alfred Webre, "NZ UFO Expert Describes Grey ET Management of

Human Souls & Bodies Aboard Grey ET Spacecraft," Exopolitics: Politics, Government, and Law in the Universe, http://bit.ly/17Y7MSq (accessed June 26, 2015); Webre, "NZ UFO Expert: On a UFO, Grey ETs merged My Son's Soul into His Body in My Womb," https://springd6.word-press.com/2014/10/16/nz-ufo-expert-on-a-ufo-grey-ets-merged-my-sons-soul-into-his-body-in-my-womb/ (accessed June 26, 2015).

2. Webre, "NZ UFO Expert Describes Grey ET Management of Human Souls"; "NZ UFO Expert: On a UFO, Grey ETs merged My Son's Soul."

3. Ibid.

4. Personal communication, statement provided to author by Suzanne Hansen via e-mail at time of interview published January 2011.

5. See "NZ UFO Expert Describes"; and "ExopoliticsTV—Grey ETs and the Human Soul with Suzanne Hansen and Alfred Lambremont," www.youtube.com/watch?v=BgAi4A2Twzw (accessed June 25, 2015).

6. Ibid.

7. Webre, "NZ UFO Expert Describes."

8. Montaldo, "Research Discussion with Host Joe Montaldo by Author on His Radio Program on 'UFO Undercover'" http://bit.ly/1aEN6v7 (accessed June 26, 2015).

Conclusion.
The Science of the Omniverse

1. "Report of Scientific Advisory Panel on Unidentified Flying Objects Convened by Office of Scientific Intelligence, CIA," http://bit.ly/15ry5Na (accessed May 15, 2015).

2. Newton, *Journey of Souls,* 123.

3. Alfred Webre, "GOOD NEWS! Universe Singularity Now Emanating Pre-Wave Energy for 'Enlightened Unity Consciousness,'" Exopolitics: Politics, Government, and Law in the Universe, http://bit.ly/1avePQ0 (accessed May 15, 2015).

4. Ibid.

5. Webre, "Is the ET Council 2015 Ecology Cleanup a Prelude to 2025 'Paradise on Earth'?" http://exopolitics.blogs.com/exopolitics/2010/12/is-the-et-council-2015-ecology-cleanup-a-prelude-to-2025-paradise-on-earth.html (accessed June 26, 2015).

6. Webre, "Mary Rodwell—ETs, Souls, The New Humans and a Coming Global Shift," http://bit.ly/17tBY7B (accessed May 13, 2015).

7. "Occupy Adam's Calendar Part 1, Extraterrestrial Genetic Manipulation," Exopolitics: Politics, Government, and Law in the Universe, http://bit.ly/W9o4jH (accessed May 15, 2015).

Bibliography

Albo, Frank. *The Cipher Golden Boy: Occult Symbolism in the Manitoba Legislative Building.* Winnipeg, Manitoba: Manitoba Culture, Heritage and Citizenship, 2010.

Allen, Rayelan. *The Obergon Chronicles.* BookSurge, 2006.

Amoroso, Richard L., and Elizabeth A. Rauscher. *The Holographic Anthropic Multiverse: Formalizing the Complex Geometry of Reality.* Oakland, Calif.: Noetic Advanced Studies Institute, Technic Research Laboratories, 2009.

Ann, Kristen. *Exploring Sacred Space.* Vast Quest Ventures, 2011.

Anthony, Sebastian. "Astronomers Estimate 100 Billion Habitable Earth-like Planets in the Milky Way, 50 Sextillion in the Universe." ExtremeTech. http://bit.ly/13UiMgU. Accessed July 1, 2015.

Barrett, Sonia. *The Holographic Canvas.* Hollywood, Calif.: Timeline Publishing, 2007.

Bateman, Wesley H. *Through Alien Eyes.* Sedona, Ariz.: Light Technology Publishing, 2000.

Belitsos, Byron. *The Adventure of Being Human: Lessons on Soulful Living from the Heart of the Urantia Revelation.* San Rafael, Calif.: Origin Press, 2012.

Berliner, Don, and Whitley Streiber, eds. "International Agreements and Resolutions—United Nations." In *UFO Briefing Document: The Best Available Evidence.* New York: Dell, 2000.

Berman, Bob, and Robert Lanza. *Biocentrism: How Life and Consciousness Are the Keys to Understanding the True Nature of the Universe.* BenBella Books, Kindle Edition, 2010.

Bertell, Rosalie. *Planet Earth: The Latest Weapon of War.* Montreal: Black Rose Books, 2001.

Blazquez, Francisca, and Joan Lluis Montane. *El Paraiso de la Nueva Luz.* Madrid, Spain: Mandala Ediciones, 2010.

Blazquez, Francisca. *Dimensionalismo.* Madrid, Spain: Museo Casa de la Moneda, 2007.

Bourdais, Gildas. *COMETA: The French Report on UFOs and Defense: A Summary.* Chicago: Center for UFO Studies (CUFOS), 1999.

Brown, Courtney. *Cosmic Explorers.* New York: Dutton, 1999.

———. *Cosmic Voyage.* Dutton, 1996.

———. *Remote Viewing: The Science and Theory of Non-Physical Perception.* Farsight Press, 2005.

Calleman, Carl Johan. *The Mayan Calendar and the Transformation of Consciousness.* Rochester, Vt.: Bear & Co., 2004.

———. *The Purposeful Universe.* Rochester, Vt.: Bear & Co., 2009.

Cameron, Grant. "UFO Studies Done and Proposed by the Carter Administration." www.presidentialufo.com, 2005. Accessed July 1, 2015.

Cameron, Tilde, and Tina Fiorde. *A Book of Insight.* Copquitlam, B.C.: Stone Circle Publishing, 2010.

Carlsberg, Kim. *Beyond My Wildest Dreams.* Santa Fe: Bear & Co., 1995.

———. *The Art of Close Encounters.* Sedona, Ariz.: Close Encounters Publishing, 2010.

Carroll, Judy. *Human by Day, Zeta by Night: A Dramatic Account of Zetas Incarnating as Humans.* Columbus, N.C.: Wildflower Press, 2011.

Carrol, Judy, and Helene Kaye. *The Zeta Message: Connecting All Beings in Oneness.* Columbus, N.C.: Wildflower Press, 2011.

Carter, Chris. *Science and Psychic Phenomena.* Rochester, Vt.: Inner Traditions, 2007.

Chalker, Bill. *Hair of the Alien: DNA and Other Forensic Evidence of Alien Abduction.* Pocket Books, 2005.

Chevalier, Andre. *I Speak to the Dead.* Montreal: Amazon.com, 2010.

Clow, Barbara Hand. *Awakening the Planetary Mind: Beyond the Trauma of the Past to a New Era of Creativity.* Rochester, Vt.: Bear & Co., 2011.

———. *Catastrophobia: The Truth Behind Earth Changes in the Coming Age of Light.* Rochester, Vt.: Bear & Co., 2001.

———. *The Mayan Code: Time Acceleration and Awakening the World Mind.* Rochester, Vt.: Bear & Co., 2007.

———. *The Mind Chronicles: A Visionary Guide into Past Lives.* Rochester, Vt.: Bear & Co., 2006.

———. *Alchemy of Nine Dimensions* (with Gerry Clow). Charlottesville, Va.: Hampton Roads, 2004.

———. *The Pleiadian Agenda: A New Cosmology for the Age of Light.* Santa Fe: Bear & Co., 1995.

Cockell, C.S. "Liberty and the Limits to the Extraterrestrial State." *J. Brit. Interplanetary Soc.* 62 (2009): 139–57.

Colli, Janet Elizabeth. *Sacred Encounters: Spiritual Awakenings during Close Encounters.* Bloomington, Ind.: Xlibiris, 2004.

Colombo, John Robert. *Extraordinary Experiences.* Toronto: Hounslow Press, 1999.

Cook, Nick *The Hunt for Zero Point.* New York: Broadway Books, 2002.

Cooper, L. Gordon. "Astronaut Gordon Cooper Addressing UN Panel Discussion on UFOs and ETs, 1985." *UFO Universe* 1, no. 3 (November 1988).

Corak, Vedia Bulent (Onsu). *The Knowledge Book.* Istanbul, Turkey: World Brotherhood Union Mevlana Supreme Foundation, 1995.

Cori, Patricia. *Before We Leave You.* Berkeley, Calif.: North Atlantic Books, 2011.

———. *No More Secrets, No More Lies.* Berkeley, Calif.: North Atlantic Books, 2008.

Corso, Philip J. *The Day After Roswell.* New York: Pocket Books, 1997.

Deane, Ashayana. *Voyagers: The Sleeping Abductees.* Columbus, N.C.: Wildfire Press, 2001.

Deardorff, James, Bernard Haisch, Bruce Maccabee, and Hal E. Puthoff. "Inflation-theory Implications for Extraterrestrial Visitation." *J. Brit. Interplanetary Soc.* 58 (2005): 43–50.

Dewhurst, Richard J. *The Ancient Giants Who Ruled America.* Rochester, Vt.: Bear & Co., 2014.

Dickinson, Terence, and Adolf Schaller. *Extraterrestrials: A Field Guide for Earthlings.* Buffalo, N.Y.: Camden House, Firefly Books, 1994.

Dolan, Richard M. *UFOs and the National Security State: Chronology of a Coverup, 1941–1973.* Newburyport, Mass.: Hampton Roads Publishing, 2002.

Dolan, Richard M., and Linda Moulton Howe. *UFOs and the National Security State, Vol. 2: The Cover-Up Exposed, 1973–1991.* Keyhole Publishing, 2009.

Dolan, Richard M., Bryce Zabel, and Jim Marrs. *A.D. After Disclosure: When the Government Finally Reveals the Truth About Alien Contact.* Wayne, N.J.: New Page Books, 2012.

Dvorsky, George. "How to Measure the Power of Alien Civilizations Using the Kardashev Scale," February 25, 2013. http://bit.ly/J3G0Yy. Accessed July 1, 2015.

Dyer, M.G. "Quantum Physics and Consciousness, Creativity, Computers: A Commentary on Goswami's Quantum-based Theory of Consciousness and Free Will." *J. Mind and Behavior* 15 (1994): 265–90.

Eisenhower, Jean. *Rattlesnake Fire.* Silver City, N.Mex.: ParadigmSalon Publishing, 2010.

Elder, Paul. *Eyes of an Angel.* Charlottesville, Va.: Hampton Roads, 2005.

Elkins, Don, Carla Rueckert, and James Allen McCarthy. *The Ra Material: An Ancient Astronaut Speaks (The Law of One, Nos. 1–IV).* Whiteford Press, 1984.

Ellegion, Michel, and Aurora Light. *Prepare for the Landings.* Scottsdale, Ariz.: Vortex, 2008.

Ellis, George F. R. "Does the Multiverse Really Exist?" Scientific American, July 19, 2011. www.scientificamerican.com/article.cfm?id=does-the-multiverse-really-exist. Accessed July 1, 2015.

Essene, Virginia, and Sheldon Nidle. *You are Becoming a Galactic Human.* Santa Clara, Calif.: Spiritual Education Endeavors, 1994.

Eure, Robert Frank. *The Mysterious Vistors.* Bloomington, Ind.: Xlibris, 2004.

Fernandes, Fernando, Joaquim Fernandes, and Raul Berenguel. *The Fatima Trilogy.* Andrew D. Basiago and Eva M. Thomson, eds. San Antonio, Tex.: Anomalist Books, 2008.

Flem-Ath, Rand, and Colin Wilson. *The Atlantis Blueprint.* New York: Time-Warner, 2000.

Fulham, Stanley A. *Challenges of Change.* Winnipeg, Manitoba: Amisk Enterprises, 2010.

Fontana, David. *Is There an Afterlife? A Comprehensive Overview of the Evidence.* Blue Ridge Summit, Pa.: NBN Books, 2005.

———. *Life Beyond Life: What Should We Expect?* London: Watkins Publishing, 2009.

Garland, Joe. "The Kardashev Scale—Type I, II, III, IV & V Civilization." Quarks to Quasars. October 8, 2013. http://bit.ly/1bNRJBm. Accessed July 1, 2015.

Garriga, Jaume, and Alexander Vilenkin. "Many Worlds in One." *Phys. Rev.* D64, no. 043511 (July 26, 2001).

Gibson, Mitchell Earl. *Your Immortal Body of Light*. Forest Hill, Calif.: Reality Entertainment, 2006.

Good, Timothy. *Alien Contact*. New York: William Morrow, 1993.

Goswami, Amit. "The Idealistic Interpretation of Quantum Mechanics." *Phys. Essays* 2 (1989): 385–400.

———. "Consciousness in Quantum Physics and the Mind-body Problem." *J. Mind and Behavior* 11, no. 1 (Winter 1990): 75–96.

———. *The Self-Aware Universe: How Consciousness Creates the Material World*. New York: Tarcher/Putnam, 1993.

———. "Monistic Idealism May Provide Better Ontology for Cognitive Science: A Reply to Dyer." *J. Mind and Behavior* 16, no. 2 (Spring 1995): 135–50.

Greene, Brian. The Hidden Reality: Parallel Universes and the Hidden Laws of the Cosmos. New York: Knopf, 2011.

Greene, Brian, et al. "What Is Life Like in Other Parts of the Multiverse?" Scientific American, November 21, 2011. www.scientificamerican.com/article.cfm?id=fabric-of-the-cosmos-multiverse. Accessed July 1, 2015.

Greer, Steven M. *Disclosure: Military and Government Witnesses Reveal the Greatest Secrets in Modern History*. Crossing Point Inc., 2001.

———. *Extraterrestrial Contact: The Evidence and Implications*. Crossing Point Inc., 2013.

———. *Hidden Truth: Forbidden Knowledge*. Crossing Point Inc., 2006.

Haitch, Elizabeth. *Initiation*. Santa Fe, N.Mex.: Aurora Press, 2000.

Hardcastle, Rebecca. *Exoconsciousness: Your 21st Century Mind*. Author House, 2008.

Harper, John Jay. *Tranceformers: Shamans of the 21ˢᵗ Century*. Forest Hill, Calif.: Reality Entertainment, 2006.

Harris, Fred, and Byron Belitsos, eds. *The Center Within*. Novato, Calif.: Origin Press, 1998.

Harris, Paola Leopizzi. *Connecting the Dots: Making Sense of the UFO Phenomenon*. Bloomington, Ind.: Author House, 2005.

———. *Exopolitics: All the Above*. Bloomington, Ind.: Author House, 2009.

———. *Exopolitics: How Does One Speak To A Ball Of Light?* Bloomington, Ind.: Author House 2007.

Hawkins, David R. *Power vs Force*. Carlsbad, Calif.: Hay House, 2002.

Haze, Xaviant. *Aliens in Ancient Egypt*. Rochester, Vt.: Bear & Co., 2013.

Hellyer, Paul. *Light at the End of the Tunnel*. Bloomington, Ind.: Author House, 2010.

Helmenstine, Anne Marie. "Scientific Hypothesis, Theory, Law definitions, Learn the Language of Science." About Education. http://bit.ly/bV90Rj. Accessed July 1, 2015.

Horowitz, Leonard. *Walk on Water*. Sandpoint, Ind.: Tetrahedron, 2006.

Huneeus, Antonio. "Statement to the Second Symposium on Extraterrestrial Intelligence and the Human Future, UN Society for Enlightenment and Transformation (SEAT)," reprinted in SEAT Newsletter, December 1993.

Hurtak, J. J. *The Keys of Enoch*. Los Gatos, Calif.: The Academy for Future Science, 1977.

Huyghe, Patrick, and Dennis Stacy. *The Field Guide to Extraterrestrials: A Complete Overview of Alien Life Forms Based on Actual Accounts and Sightings*. New York: Avon Books, 1996.

Hynek, J. Allen. "Dr. J. Allen Hynek Speaking at the United Nations, November 27th, 1978." UFO Evidence.com. www.ufoevidence.org/documents/doc757 .htm. Accessed July 1, 2015.

———. *The UFO Experience: A Scientific Inquiry*. Washington D.C.: Marlowe & Co., 1998.

Hynek, J. Allen, and Jacques Vallée. *The Edge of Reality: A Progress Report on the Unidentified Flying Objects*. Oregon, Ill.: Quality Books, 1975.

Jacobs, David M. *The Threat*. New York: Simon & Schuster, 1998.

Julien, Eric. *The Science of Extraterrestrials*. Fort Oglethorpe, Ga.: Allies Books, 2006.

Jung, Carl G. "Dr. Carl Jung on Unidentified Flying Objects." *Flying Saucer Review* 1, no. 2 (1955).

Justice, J. *DNA in the Sands of Time*. Pittsburgh, Pa.: Rosedog Books, 2006.

Kanazawa, Satoshi. "Do Extraordinary Claims Require Extraordinary Evidence?" Psychology Today. http://bit.ly/1chFcqK. Accessed July 1, 2015.

Kardashev, Nikolai. "Transmission of Information by Extraterrestrial Civilizations." *Soviet Astron* 8 (1964): 217.

Kardec, Allan. *The Spirits Book: 1019 Questions and Answers about the Immortality of the Soul*. Guilford, U.K.: White Crow, 2009.

Kean, Leslie. *UFOs: Generals, Pilots, and Government Officials Go on the Record*. New York: Three Rivers Press, 2011.

Kitei, Lynne D. *The Phoenix Lights*. Charlottesville, Va.: Hampton Roads, 2004.

Kling, Peter. *Letters to Earth*. Durham, Conn.: Eloquent Books, 2009.

Knight-Jadczyk, Laura. *High Strangeness: Hyperdimensions and the Process of Alien Abduction*. Grand Prairie, Alberta: Red Pill Press, 2008.

Komarek, Ed. *UFOs, Exopolitics, and the New World Disorder*. Cairo, Ga.: Shoestring Publishing, 2012.

Kramer, Miriam. "Scale of Universe Measured with 1 Percent Accuracy." Space. com. http://bit.ly/1e96Rhq. Accessed July 1, 2015.

Krapf, Philip H. *The Contact Has Begun*. Carlsbad, Calif.: Hay House, 2001.

———. *Meetings with Paul: An Atheist Discovers His Guardian Angel*. San Rafael, Calif.: Origin Press, 2008.

Krauss, Lawrence M. *A Universe from Nothing*. New York: Simon & Schuster, 2012.

Kubris, P., and M. Macy. *Conversations Beyond the Light with Departed Friends and Colleagues by Electronic Means*. Boulder, Colo.: Griffin Publishing, 1995.

LaViolette, Paul A. *Decoding the Message of the Pulsars*. Rochester, Vt.: Bear & Co., 2006.

———. *Secrets of Antigravity Propulsion*. Rochester, Vt.: Bear & Co., 2008.

———. *The Talk of the Galaxy*. Alexandria, Va.: Starlane Publications, 2000.

Ledwith, Miceal, and Klaus Heinemann. *The Orb Project*. New York: Atria Books, 2007.

Linde, Andrei. "Inflation, quantum cosmology, and the anthropic principle." In J. D. Barrow, P. C. W. Davies, and C. L. Harper, eds., *Science and Ultimate Reality: From Quantum to Cosmos*. Cambridge: Cambridge University Press, 2003.

Lipton, Bruce H. *The Biology of Belief*. Carlsbad, Calif.: Hay House, 2008.

Lloyd, Andy. *Dark Star*. Santa Barbara, Calif.: Timeless Voyager Press, 2005.

Lobuono, George. *Alien Mind: The Thought and Behavior of Extraterrestrials*. Davis, Calif.: QC Press, 2010.

Lora, Doris, and Russell Targ. "How I Was a Psychic Spy for the CIA and Found God." Okland, Calif.: Institute for Noetic Sciences (IONS), 2003.

Lorgen, Eve. *The Love Bite: Alien Interference in Human Love Relationships*. Bonsall, Calif.: E Logos, 1999.

Love, Frank. *The Broken Code*. Bloomington, Ind.: Author House, 2007.

Mack, John E. *Passport to the Cosmos: Human Transformation and Alien Encounters*. New York: Crown, 1999.

Maddaloni, Vincenzo, "Invention of a Machine for Photographing the Past." *Domenica del Corriere* 74, no.18 (May 2, 1972): 26–29.

Manning, Jeane, and Joel Garbon. *Breakthrough Power.* Vancouver, B.C.: Ambre Bridge, 2009.

Mannion, Michael. *Project Mindshift: The Re-education of the American Public Concerning Extraterrestrial Life.* M Evans and Co., 1998.

Markowitz, Clara. "Five Reasons We May Live in a Multiverse." Space.com, December 7, 2012. http://bit.ly/18Q7t8N. Accessed July 1, 2015.

Marrs, Jim. *Alien Agenda: Investigating the Extraterrestrial Presence Among Us.* New York: Harper Collins, 2000.

———. *Psi Spies.* Phoenix, Ariz.: Alien Zoo Publishers, 2000.

Marzulli, L.A. *Politics, Prophecy & the Supernatural.* Crane, Mo.: Anomalos Publishing, 2007.

McDonald, James E. "Dr. James McDonald's letter and statement to the United Nations, June 7th, 1967." UFO Evidence.com. www.ufoevidence.org/documents/doc1056.htm. Accessed July 1, 2015.

Meier, Eduard. *UFOs: Spaceships from Foreign Worlds.* Schmidrüti, Switzerland: FIGU, 1999.

Miejan, Tim. "Journey of Souls with Michael Newton." Edge Magazine, July 1997. http://bit.ly/1bpsY2m. Accessed July 1, 2015.

Moody, Raymond A., Jr. *Life after Life.* San Francisco: Harper One, 1975.

Moore, Judith K. *Song of Freedom.* Flagstaff: Light Technology Publishing, 2002.

Moroney, Jim. *The New Bridge: Planning for the Extraterrestrial Presence.* Calgary: SCMC, 2007.

Munoz, Daniel. *Enigma de los Circulos de Inglaterra.* Jiutepec, Morelos, Mexico: Espacio Alternativo, 2009.

Newlove-Eriksson, Lindy, and Eriksson, Johan. "Governance Beyond the Global: Who Controls the Extraterrestrial?" *Globalizations* 10, no. 2 (2013).

Newton, Michael. *Life Between Lives: Hypnotherapy for Spiritual Regression.* Woodbury, Minn.: Llewellyn Publications, 2004.

———. *Destiny of Souls: New Case Studies of Life Between Lives.* Woodbury, Minn.: Llewellyn, 2009.

———. *Journey of Souls: Case Studies of Life Between Lives.* Woodbury, Minn.: Llewellyn, 2008.

Norman, Ernest L. *The Truth About Mars.* El Cajon, Calif.: Unarius Academy of Science, 1998.

Oakford, David L. *Journey through the World of Spirit, God, Gaia, and Guardian Angels*. Forest Hill, Calif.: Reality Entertainment, 2007.

Ocean, Joan. *Dolphin Connection*. Kailua, Hawaii: Dolphin Connection, 1996.

Ostrander, Sheila, and Lynn Schroeder. *Psychic Discoveries Behind the Iron Curtain*. New York: Random House, 1971.

Palacios, Rafael. *Extraterrestres: El Secreto Mejor Guardado*. Madrid, Spain: Palmyra, 2009.

Paquin, Marguerite. *Manual for the Soul*. White Pup Press, 2009.

Peake, Anthony. *Is There Life After Death?* London: Arcturus Publishing, 2006.

Pearse, Steve. *Set Your Phaser to Stun*. Bloomington, Ind.: Xlibris, 2011.

Penrose, Roger. *The Emperor's New Mind*. Oxford: Oxford University Press, 1989.

———. *Shadows of the Mind*. Oxford: Oxford University Press, 1994.

Penrose, Roger, and S.R. Hameroff. "Orchestrated Reduction of Quantum Coherence in Brain Microtubules: A Model for Consciousness." In S. R. Hameroff, A. W. Kaszniak, and A. C. Scott, eds., *Toward a Science of Consciousness: The First Tucson Discussions and Debates*. Cambridge: MIT Press, 1996.

Pereira, Patricia. *Arcturan Star Chronicles*. Hillsboro, Ore.: Beyond Words Publishing, 1998.

Pinchbeck, Daniel. *2012: The Return of Questzalcoatl*. New York: Tarcher/Penguin, 2006.

"Possibilities in Parallel: Seeking the Multiverse." Scientific American, May 20, 2013. http://books.scientificamerican.com/sa-ebooks/books/possibilities-in-parallel-seeking-the-multiverse/. Accessed July 1, 2015.

Proud, Louis. *The Secret Influence of the Moon*. Rochester, Vt.: Destiny Books, 2013.

Puthoff, Hal E. "CIA-Initiated Remote Viewing at Stanford Research Institute." *The Intelligencer: Journal of US Intelligence Studies* 12, no. 1 (2001).

Rees, Martin. *Our Cosmic Habitat*. Princeton, N.J.: Princeton University Press, 2001.

Religa, Stella, and Byron Belitsos. *The Secret Revelation: Unveiling the Mystery of the Book of Revelation*. Novato, Calif.: Origin Press, 2002.

Reynolds, C. F. *Invoking the Light*. Maryborough, Victoria, Australia: McPherson's Printing Group, 2007.

Rhine, J. B. *Extra-Sensory Perception*. Boston: Bruce Humphries, 1934.

———. *New World of the Mind*. New York: William Sloane Associates, 1953.

———, ed. *Progress in Parapsychology*. Durham, N.C.: Parapsychology Press, 1971.

Rhine, J. B., and R. Brier, eds. *Parapsychology Today*. New York: Citadel Press, 1968.

Rhine, J. B., and J. G. Pratt. *Parapsychology: Frontier Science of the Mind*. Springfield, Ill.: Charles C. Thomas, 1957.

Ring, Kenneth. *Lessons from the Light: What We Can Learn from Near-Death Experiences*. Needham, Mass.: Moment Point Press, 1998, 2006.

Risi, Armin. *TranscEnding the Global Power Game*. Flagstaff: Light Technology Publishing, 2003.

Robinson, James, et al., eds. and trans. *The Nag Hammadi Library in English*. New York: Harper, 1989.

Rosen-Bizberg, Franz. *Orion Transmissions Prophecy, Vol. I*. Poland: Fundacja Terapia Homa, 2003.

Salla, Michael E. *Exopolitics: Political Implications of the Extraterrestrial Presence*. Tempe, Ariz.: Dandelion Books, 2003.

———. *Exposing U.S. Government Policies on Extraterrestrial Life: The Challenge of Exopolitics*. Pahoa, Hawaii: Exopolitics Institute, 2012.

———. *Galactic Diplomacy: Getting to Yes with ET*. Pahoa, Hawaii: Exopolitics Institute, 2013.

———. *Kennedy's Last Stand: Eisenhower, UFOs, MJ-12 & JFK's Assassination*. Pahoa, Hawaii: Exopolitics Institute, 2013.

Sambhava, Padma, and Robert Thurman. *The Tibetan Book of the Dead*. New York: Viking, 2006.

Sauder, Richard. *The Richard Sauder Briefing*. Roman Schmitt Verlag, 2010.

Schuessler, John F. "UFOs, the UN, and GA 33/426." HUFON Report Newsletter, December 1992. www.ufoevidence.org/documents/doc748.htm. Accessed July 1, 2015.

Scott, A. *Stairway to the Mind: The Controversial New Science of Consciousness*. New York: Springer-Verlag, 1995.

Seife, Charles. *Zero: The Biography of a Dangerous Idea*. London: Creative Print & Design, 2000.

Silva, Freddy. *Secrets of the Fields: The Science and Mysticism of Crop Circles*. Hampton Roads, Va.: Hampton Roads Publishing Co., 2002.

Sivananda, Swami. *Mind: Its Mysteries and Control*. Divine Life Society, 2009.

Smith, Jerry E., and George Piccard. *Secrets of the Holy Lance*. Kempton, Ill.: Adventures Unlimited Press, 2005.

Sparks, Jim. *The Keepers*. Columbus, N.C.: Wildflower Press, 2006.

Stanley, Robert M. *Covert Encounters in Washington, DC*. Providence, R.I.: Unicus Press, 2011.

Stapp, Henry. "The Copenhagen Interpretation." *Amer. J. Phys.* 40 (1972): 1098–116.

———. *Mind, Matter, and Quantum Mechanics*. New York: Springer, 1993.

———. "Attention, Intention, and Will in Quantum Physics." *J. Consc. Stud.* 6 (May 24, 1999).

Steiger, Brad. *Guardian Angels and Spirit Guides*. New York: Signet, 1995.

Steinberg, Max. *Celestial Science*. New York: Steinberg, 2011.

Stevenson, Ian. "A New Look at Maternal Impressions: An Analysis of 50 Published Cases and Reports of Two Recent Examples." *J. Scientific Exploration* 6 (1992): 353–373.

———. "Phobias in Children Who Claim to Remember Previous Lives." *J. Scientific Exploration* 4 (1990): 243–54.

———. "American Children Who Claim to Remember Previous Lives." J. *Nervous and Mental Disease* 17, no. 1 (1983): 742–48.

———. *Cases of the Reincarnation Type, I: Ten Cases in India*. Charlottesville: University Press of Virginia, 1975.

———. *Children Who Remember Previous Lives*. Charlottesville: University Press of Virginia, 1987.

———. *Science, the Self, and Survival After Death: Selected Writings*. Emily Williams Kelly, ed. Rowman and Littlefield, 2013.

Streiber, Whitley. *Transformation: The Breakthrough*. New York: William Morrow, Beech Tree Books, 1988.

Sturrock, Peter A. *The UFO Enigma*. New York: Warner Books, 1999.

Susskind, Leonard. *The Cosmic Landscape: String Theory and the Illusion of Intelligent Design*. New York: Back Bay Books, 2006.

Talbot, Michael. *The Holographic Universe*. New York: Harper Perennial, 1991.

Targ, Russell. *The Reality of ESP: A Physicist's Proof of Psychic Abilities*. Wheaton, Ill.: Quest Books, 2012.

Targ, Russell, and Harold Puthoff. *Mind-Reach*. Charlottesville, Va.: Hampton Roads, 1977.

Targ, Russell, and Katra, Jane. "The Scientific and Spiritual Implications of Psychic Abilities." *Altern Ther Health Med.* 7, no. 3 (2001): 143–49.

Tegmark, Max. "Is 'The Theory of Everything' Merely the Ultimate Ensemble Theory?" *Annals Phys.* 270, no.1 (November 20, 1998): 1–51.

Tellinger, Michael, and Johan Heine. *Temples of the African Gods.* Waterval Boven, South Africa: Zulu Planet, 2009.

Tellinger, Michael. *Slave Species of God.* Johannesburg, South Africa: Zulu Planet, 2009.

Torres, Noe, and Ruben Uriarte. *Mexico's Roswell: The Chihuahua UFO Crash.* Roswell Books, 2008.

Tramont, C. V. *From Birth to Rebirth: Gnostic Healing for the 21st Century.* Columbus, N.C.: Granite Publishing, 2008.

Tremblay, Rodrigue. *The Code for Global Ethics.* Amherst, N.Y.: Prometheus Books, 2010.

Tunich, Jose Alberto. *Encuentros Cercanos con Extraterrestres.* Buenos Aires, Argentina: CEFAE, 1999.

Vasquez, John, and Bruce Stephen Holms. *Incident at Fort Benning.* Santa Barbara, Calif.: Timeless Voyager Press, 2000.

Vaughan-Lee, Llewellyn. *Awakening the World.* Inverness, Calif.: Golden Sufi Center, 2006.

Vilenkin, Alexander. Many Worlds in One: The Search for Other Universes. New York: Hill and Wang, 2006.

Vilenkin, Alexander, and Max Tegmark. "The Case for Parallel Universes." Scientific American, July 11, 2011. www.scientificamerican.com/article. cfm?id=multiverse-the-case-for-parallel-universe. Accessed July 1, 2015.

Von Neumann, John. *Mathematical Foundations of Quantum Mechanics.* Princeton, N.J.: Princeton University Press, 1955.

Wallace, Tim. *Hidden Wisdom.* Newburyport, Mass.: Disinformation Books, 2010.

Ward, Suzanne. *Matthew: Tell Me About Heaven: A Firsthand Description of the Afterlife.* Camas, Wash.: Matthew Books, 2006.

Webre, Alfred Lambremont. *Exopolitics: A Decade of Contact.* Vancouver, B.C.: Universe Books, 2000.

———. *Exopolitics: Politics, Government, and Law in the Universe.* Vancouver, B.C.: Universe Books, 2005.

———. *The Age of Cataclysm.* New York: Berkley Pub Corp, 1975.

———. *The Levesque Cases.* Ontario: PSP Books, 1990.

Wilber, Ken. *A Brief History of Everything.* Boston, Mass.: Shamballa, 2007.

Wilber, Ken. *Integral Spirituality: A Startling New Role for Religion in the Modern and Post Modern World*. Boston, Mass.: Integral Books, 2007.

Wilcock, David. *The Source Field Investigations: The Hidden Science and Lost Civilizations Behind the 2012 Prophecies*. Brilliance Audio, 2012.

———. *The Synchronicity Key: The Hidden Intelligence Guiding the Universe and You*. Penguin Group US Kindle Edition, 2013.

Wilson, Simon Harvey. "Shamanism and Alien Abductions: A Comparative Study." *Australian J. Parapsychol.* 1, no. 2 (December 2001).

Wolchover, Natalie. "New Physics Complications Lend Support to Multiverse Hypothesis." Scientific American, June 1, 2013. www.scientificamerican. com/article.cfm?id=new-physics-complications-lend-support-to-multiverse-hypothesis. Accessed July 1, 2015.

Wood, Ryan S. *Majic Eyes Only: Earth's Encounters with Extraterrestrial Technology*. Wood Enterprises, 2005.

Woolf, Victor Vernon. *The Dance of Life*. Reno, Nev.: International Academy of Holodynamics, 2005.

Wyllie, Timothy. *Revolt of the Rebel Angels: The Future of the Multiverse*. Rochester, Vt.: Bear & Co., 2013.

Zammit, Victor, and Wendy Zammit. *A Lawyer Presents the Evidence for the Afterlife*. Hove, U.K.: White Crow Books, 2013.

Index

About the Author

In 2000 Alfred Lambremont Webre founded exopolitics, the science of relations among intelligent civilizations in the multiverse, when he published his book *Exopolitics: A Decade of Contact.* Using advanced quantum access technology, the U.S. Defense Advanced Research Projects Agency (DARPA) and the Central Intelligence Agency (CIA) time traveled the 2005 edition, *Exopolitics: Politics, Government, and Law in the Universe,* back to 1971, when an unwitting Alfred was examined by a group of approximately fifty CIA and DARPA officials who knew he would be a leading future extraterrestrial and time-travel whistle-blower and would become the developer of exopolitics and the dimensional ecology hypothesis of the Omniverse.

Here, Alfred integrates empirical data from intelligent extraterrestrial civilizations and from the intelligent civilizations of souls in the interlife and demonstrates a new hypothesis of a functioning ecology of intelligence in the dimensions in the Omniverse.

Alfred is a graduate of Yale University and Yale Law School in international law and was a Fulbright Scholar in international economic integration in Uruguay. He has taught economics at Yale University and constitutional law at the University of Texas. Alfred was general counsel to the New York City Environmental Protection Administration, a futurist at Stanford Research Institute (where he directed the proposed 1977 Carter White House extraterrestrial communication study), and was an NGO delegate to the United Nations and the UNISPACE conference. He has been a judge on the Kuala Lumpur War Crimes Tribunal.

Alfred is active with nonprofit organizations concerned with peace, the environment, human rights and social justice, consciousness, freedom from neural and mind control, and intelligent life in the Omniverse.

Alfred may be contacted through

www.exopolitics.com

Books of Related Interest

Secrets of Antigravity Propulsion
Tesla, UFOs, and Classified Aerospace Technology
by Paul A. LaViolette, Ph.D.

Secret Journey to Planet Serpo
A True Story of Interplanetary Travel
by Len Kasten

The Secret History of Extraterrestrials
Advanced Technology and the Coming New Race
by Len Kasten

The Secret Influence of the Moon
Alien Origins and Occult Powers
by Louis Proud

The John Michell Reader
Writings and Rants of a Radical Traditionalist
by John Michell
Introduction by Joscelyn Godwin

Grey Aliens and the Harvesting of Souls
The Conspiracy to Genetically Tamper with Humanity
by Nigel Kerner

Extra-Planetary Experiences
Alien-Human Contact and the Expansion of Consciousness
by Thomas James Streicher, Ph.D.
Foreword by Stanley Krippner, Ph.D.

Dolphins, ETs & Angels
Adventures Among Spiritual Intelligences
by Timothy Wyllie

INNER TRADITIONS • BEAR & COMPANY
P.O. Box 388
Rochester, VT 05767
1-800-246-8648
www.InnerTraditions.com

Or contact your local bookseller